Current and Future Cellular Systems

IEEE Press
445 Hoes Lane
Piscataway, NJ 08854

# Current and Future Cellular Systems

## Technologies, Applications, and Challenges

*Edited by*

*Garima Chopra*
Chitkara University Institute of Engineering and Technology
Chitkara University, Punjab, India

*Suhaib Ahmed*
Model Institute of Engineering and Technology
Jammu, J&K, India

*Shalli Rani*
Chitkara University Institute of Engineering and Technology
Chitkara University, Punjab, India

IEEE Press
WILEY

Published by John Wiley & Sons, Inc., Hoboken, New Jersey.
Published simultaneously in Canada.



***Library of Congress Cataloging-in-Publication Data Applied for:***

**Hardback ISBN: 9781394256044**

Cover Design: Wiley
Cover Image: © WD Ashari/Shutterstock

Set in 9.5/12.5pt STIXTwoText by Straive, Chennai, India

# Contents

# About the Editors

**Garima Chopra**

Dr. Garima Chopra is currently working as an assistant professor, Department of Electronics and Communication, Chitkara University, Punjab. She received the B Tech degree in Electronics and Communication from Shri Mata Vaishno Devi University, Jammu and Kashmir, India, in 2013 and MTech degree specialized in Communication and Information Technology (CIT) in Electronics and Communication Department from National Institute of Technology, Srinagar, Jammu and Kashmir, India, in 2016. She completed her PhD in Electronics and Communication Engineering at Shri Mata Vaishno Devi University, Jammu and Kashmir, India. She was a postdoctoral fellow at IIT Hyderabad from 2021 to 2022 (SERB-NPDF). Her research interest includes the security in emerging technologies of 5G wireless communication network. Currently she is doing her research work on security issues of Ultra Dense Network in Fifth Generation wireless communication. She also worked on the security issues of wireless communication and their security aspects.

**Suhaib Ahmed**

Dr. Suhaib Ahmed is currently working as an assistant professor and faculty-in-charge, Centre for Research, Innovation and Entrepreneurship, Model Institute of Engineering and Technology, Jammu, India. He received his BE degree in Electronics and Communication Engineering from University of Jammu, India, in 2012, MTech and PhD degrees in Electronics and Communication Engineering from Shri Mata Vaishno Devi University, Katra, India, in 2014 and 2019, respectively. His research interests include quantum-dot cellular automata, nano-electronics, VLSI, data converters, nanotechnology, wireless sensor networks, IoT, AI/ML, bioelectronics, and signal processing. Dr. Ahmed is a Senior Member of IEEE, Life Member of IETE, ISTE, and Computer Society of India. He has published around 90 research papers in reputed peer-reviewed journals of IEEE, Springer, Elsevier, Wiley, etc. and in various international/national conferences. He has also worked as PI/Co-PI in 5 sponsored projects from the Ministry of Education, Ministry of Communications, Government of India and Jammu & Kashmir Department of Science and Technology. He has also served as TPC in various conferences and served as reviewer in various journals/conferences. He is also a member of editorial boards for various Web of Science and SCOPUS indexed journals including Scientific Reports, Plos One, IET Quantum Communications, etc.

**Shalli Rani**

Dr. Shalli Rani completed her post doctorate from Manchester Metropolitan University, UK, in June, 2023. She is professor in Chitkara University Institute of Engineering and Technology, Chitkara University, Rajpura, Punjab, India. She has 18+ years of teaching experience. She received MCA degree from Maharishi Dayanand University, Rohtak, in 2004 and the MTech degree in Computer Science from Janardan Rai Nagar Vidyapeeth University, Udaipur, in 2007, and PhD degree in Computer Applications from Punjab Technical University, Jalandhar, in 2017. Her main area of interest and research are Wireless Sensor Networks, Underwater Sensor networks, Machine Learning, and Internet of Things. She has published/accepted/presented more than 100+ papers in international journals/conferences (SCI+Scopus) and edited/authored five books with international publishers. She is serving as the associate editor of *IEEE Future Directions Letters*. She served as a guest editor in *IEEE Transaction on Industrial Informatics*, Hindawi, WCMC, and Elsevier IoT Journals. She has also served as reviewer in many repudiated journals of IEEE, Springer, Elsevier, IET, Hindawi, and Wiley. She has worked on Big Data, Underwater Acoustic Sensors, and IoT to show the importance of WSN in IoT applications. She received a Young Scientist Award in February 2014 from Punjab Science Congress, Lifetime Achievement Award, and Supervisor of the Year Award from Global Innovation and Excellence, 2021.

# List of Contributors

*Laith Abualigah*
MEU Research Unit
Middle East University
Amman
Jordan

*Aditya Soni*
Chitkara University Institute of Engineering and Technology
Chitkara University
Rajpura
Punjab
India

*Yasir Ahmad*
Department of Information Technology & Security
College of Computer Science & Information Technology
Jazan University
Jazan
Kingdom of Saudi Arabia

*Suhaib Ahmed*
Department of Electronics and Communication Engineering
Model Institute of Engineering and Technology
Jammu
J&K
India

*Sherif Tawfik Amin*
Department of Computer Science
College of Computer Science & Information Technology
Jazan University
Jazan
Kingdom of Saudi Arabia

*Himanshi Babbar*
Chitkara University Institute of Engineering and Technology
Chitkara University
Rajpura
Punjab
India

***Durga Shankar Baggam***
Department of CSE
Gandhi Engineering College
Gandhi Vihar
Khurda
Odisha
India

and

Chitkara University Institute of Engineering and Technology
Chitkara University
Rajpura
Punjab
India

***Aditya Bakshi***
School of Technology Management and Engineering
Narsee Monjee Institute of Management Sciences
Chandigarh
India

***Shailesh Pramod Bendale***
Department of Computer Engineering
NBN Sinhgad School of Engineering
Pune
Maharashtra
India

***Abhijit Chitre***
Department Electronics and Telecommunication Engineering
Vishwakarma Institute of Information Technology
Pune
Maharashtra
India

***Garima Chopra***
Chitkara Institute of Engineering and Technology
Chitkara University
Rajpura
Punjab
India

***Araddhana Arvind Deshmukh***
Department of Computer Science & Information Technology (Cyber Security)
Symbiosis Skill and Professional University
Kiwale
Pune
Maharashtra
India

***Amol Dhumane***
Computer Science and Engineering Department
Symbiosis Institute of Technology
Lavale
Pune
Maharashtra
India

***Ekta Dixit***
Department of Computer Science
S.S.D Women's Institute of Technology
Bathinda
Punjab
India

***Geetanshi***
Chitkara University Institute of Engineering and Technology
Chitkara University
Rajpura
Punjab
India

***Akhil Gupta***
School of Electronics & Electrical Engineering
Lovely Professional Engineering
Phagwara
Punjab
India

***Mohammad Alamgir Hossain***
Department of Computer Science
College of Computer Science & Information Technology
Jazan University
Jazan
Kingdom of Saudi Arabia

***Sheela Hundekari***
MITCOM, MCA Department
MIT ADT University
Loni Kalbhor
Pune
Maharashtra
India

***Brejesh Lall***
Bharti School of Telecommunication Technology and Management
IIT Delhi
New Delhi
India

***Suresh Limkar***
Department of Artificial Intelligence & Data Science
AISSMS Institute of Information Technology
Pune
Maharashtra
India

***Suresh Limkar***
Department of Artificial Intelligence & Data Science
AISSMS Institute of Information Technology
Pune
Maharashtra
India

***Cherry Mangla***
Assistant Professor
Concordia University of Edmonton
Edmonton
AB
Canada

***Harshit Manocha***
Chitkara University Institute of Engineering and Technology
Chitkara University
Punjab
India

***Arushi Pandey***
School of Technology Management and Engineering
Narsee Monjee Institute of Management Sciences
Chandigarh
India

***Ankita Rana***
Chitkara University Institute of Engineering and Technology
Chitkara University
Rajpura
Punjab
India

*Shalli Rani*
Chitkara University Institute of Engineering and Technology
Chitkara University
Rajpura
Punjab
India

*Shalli Rani*
Chitkara University Institute of Engineering and Technology
Chitkara University
Rajpura
Punjab
India

*Kanica Sachdev*
Bharti School of Telecommunication Technology and Management
IIT Delhi
New Delhi
India

*Ankita Sharma*
Chitkara University Institute of Engineering and Technology
Chitkara University
Rajpura
Punjab
India

*Sheetal Sharma*
IT- Department
Goswami Ganesh Dutta Santana
Dharma College
Chandigarh
India

*Shishir Shrivastava*
Chitkara University Institute of Engineering and Technology
Chitkara University
Rajpura
Punjab
India

*Shruti*
PG Depatment of Information Technology
Goswami Ganesh Dutta Sanatan
Dharma College
Chandigarh
India

and

Chitkara University Institute of Engineering and Technology
Chitkara University
Rajpura
Punjab
India

*Ashu Taneja*
Chitkara University Institute of Engineering and Technology
Chitkara University
Rajpura
Punjab
India

***Neeti Taneja***
Department of Computer Science Engineering
Noida Institute of Engineering and Technology
Greater Noida
Uttar Pradesh
India

***Kirti Wanjale***
Department Computer Engineering
Vishwakarma Institute of Information Technology
Pune
Maharashtra
India

# Preface

The explosive demand for high data rates by users, coupled with the rapid development of data-intensive applications, has pushed the boundaries of current cellular technologies. The evolution from 1G to 5G has been transformative, but it has also revealed the limitations of these systems in addressing the diverse needs of future networks. As we approach the era of 5G and beyond, there is a growing realization that these technologies are no longer confined to a single domain. They encompass a wide range of applications, each with unique requirements, bringing about new challenges that demand innovative solutions.

This book provides a comprehensive examination of emerging technologies and the challenges faced in the deployment of 5G and beyond. It highlights the latest trends in resource allocation, power optimization, and the role of artificial intelligence (AI) and machine learning (ML) in enhancing quality of service (QoS) and decision-making processes in complex systems such as autonomous vehicles, resource allocation, and power management. The content aims to address the architectural and deployment challenges posed by the increasing heterogeneity of devices and user demands, offering insights into how future networks can evolve to meet these requirements.

The scope of this book spans several key areas, beginning with an overview of the evolution of cellular communication, technological advancements from 1G to 5G, and the critical architectural challenges that remain. A detailed discussion on resource allocation and power optimization for 5G and beyond is provided, exploring techniques and solutions that address the growing complexity of modern cellular networks. Additionally, the growing role of AI and ML in 5G networks is examined, showcasing how these technologies can assist in areas like resource assignment, power optimization, and autonomous decision-making.

Furthermore, the book investigates into the cutting-edge domain of millimeter wave (mmWave) communication, outlining the latest trends and challenges in high-frequency communication, i.e. Terahertz frequencies (THz), systems. Moreover, it also discusses the evolving technologies and highlights the new applications in the 5G domain. This book not only provides a clear understanding of the current state of 5G technology but also inspires future research and innovation. It is our hope that this book will serve as a guide for both seasoned professionals and newcomers to the field, helping them navigate the complexities of future networks and contribute to the advancement of next-generation communication technologies.

# Glossary

- **5G Network:** The fifth generation of mobile network technology offering enhanced speed, reliability, and connectivity for applications such as IoT, autonomous vehicles, and smart cities.
- **6G Network:** The anticipated sixth generation of wireless communication technology, expected to provide ultra-high data rates, low latency, and massive device connectivity by 2030.
- **IoT (Internet of Things):** A network of interconnected devices that communicate and exchange data autonomously, often used in industries such as healthcare, transportation, and home automation.
- **Edge Computing:** A computing paradigm that processes data near its source to reduce latency, improve efficiency, and enhance real-time decision-making capabilities.
- **Machine Learning (ML):** A branch of artificial intelligence that enables systems to learn and improve from data without explicit programming, used for predictive analytics and decision-making.
- **Deep Learning (DL):** A subset of machine learning involving neural networks with many layers, used for tasks such as intrusion detection and image recognition.
- **Spectrum Sharing:** A technique for efficient use of radio frequency spectrum by allowing multiple users or devices to share the same frequency bands dynamically.
- **Cognitive Radio:** A radio technology that intelligently detects unused spectrum and adjusts transmission parameters to optimize utilization without interfering with primary users.
- **Handover (HO):** A process in wireless communication where a mobile device switches connection from one base station to another to maintain seamless connectivity.
- **Distributed Denial of Service (DDoS):** A cyberattack where multiple systems overwhelm a target, such as a server or network, causing disruption of services.

- **Network Slicing:** A 5G feature that creates multiple virtual networks on a shared physical infrastructure to cater to specific use cases with different requirements.
- **Artificial Intelligence (AI):** The simulation of human intelligence processes by machines, especially computer systems, enabling tasks like learning, reasoning, and self-correction.
- **Cybersecurity:** The practice of protecting systems, networks, and data from digital attacks, unauthorized access, and damage.
- **Blockchain:** A decentralized digital ledger technology used for secure and transparent data recording and verification across distributed networks.
- **Non-orthogonal Multiple Access (NOMA):** A multiple access technique allowing multiple users to share the same resources (frequency/time) by allocating different power levels to optimize spectral efficiency.
- **Interference Mitigation:** Strategies and techniques to reduce or eliminate signal interference in wireless communication systems to improve performance.
- **Dynamic Spectrum Access (DSA):** A method that allows unlicensed users to access spectrum dynamically based on availability and usage patterns.

# Introduction

The surging demand for high data rates and the rapid proliferation of data-intensive applications have pushed the boundaries of current cellular technologies. The evolution from the first-generation (1G) to fifth-generation (5G) mobile networks has been transformative, enabling faster speeds, lower latency, and broader connectivity. However, these advancements have also highlighted the limitations of existing systems in addressing the diverse needs of future networks. As we enter the 5G era and beyond, these technologies are no longer confined to a single domain; instead, they encompass a wide array of applications, each with unique requirements. This shift introduces unprecedented challenges that necessitate innovative solutions to meet the demands of an increasingly interconnected world.

This book delves into the emerging technologies and challenges shaping the deployment of 5G and beyond, offering a comprehensive examination of how these advancements can overcome the complexities of modern cellular systems. It addresses critical aspects of resource allocation, power optimization, and the integration of Artificial Intelligence (AI) and Machine Learning (ML) to enhance decision-making processes and Quality of Service (QoS). These technologies have proven instrumental in addressing complex problems in areas like autonomous systems, power management, and resource assignment.

This book is structured to provide readers with a clear understanding of the evolving technological landscape. It begins by tracing the evolution of cellular communication from 1G to 5G, highlighting the critical architectural challenges that persist. The discussion then delves into resource allocation and power optimization techniques, essential for managing the increasing complexity of modern networks.

A significant portion of this book explores the role of AI and ML in 5G networks, showcasing their potential to revolutionize resource management and autonomous decision-making. The book also delves into cutting-edge domains like millimeter wave (mmWave) and terahertz (THz) communication, highlighting the latest trends and challenges associated with high-frequency systems. Furthermore, it investigates emerging applications within the 5G domain, demonstrating how they are reshaping industries and enabling innovative use cases.

This book aims to not only provide a clear understanding of the current state of 5G technology but also to inspire future research and innovation in the field. By offering insights into the architectural and deployment challenges posed by the heterogeneity of devices and user demands, it seeks to guide both seasoned professionals and newcomers in navigating the complexities of future networks.

We hope this book serves as a valuable resource for those seeking to understand and contribute to the advancement of next-generation communication technologies, fostering a future of seamless, intelligent, and interconnected networks.

# 1

# Spectrum Sharing Schemes for 5G and Beyond in Wireless Communication

*Aditya Bakshi[1], Akhil Gupta[2], and Arushi Pandey[1]*

[1] *School of Technology Management and Engineering, Narsee Monjee Institute of Management Studies, Chandigarh, India*
[2] *School of Electronics & Electrical Engineering, Lovely Professional Engineering, Phagwara, Punjab, India*

## 1.1 Introduction

In last 10 years, a great advancement has been done in increasing the number of wireless broadband and multimedia devices after evolution of the first-generation (1G) mobile communication system. New services and use cases have evolved using fifth-generation (5G) wireless networks, namely, enhanced mobile broadband (eMBB), ultra-reliable low latency communications (URLLC) etc. The eMBB shows enhancements in user data rates whereas URLLC is used for mission-critical and time-sensitive applications which provide low-latency and ultra-high reliability. Enabling communication between a great number of devices can be achieved by massive machine-type communications (m-MTC). The evolution of the Internet of Things (IoT) and its subset, industrial IoT (IIoT) helps to generate efficient and sustainable production using ubiquitous Internet connections [1]. Beyond 5G, the next i.e. sixth-generation (6G) communication systems plays an important role in accelerating the IOT implementation and diverse industry areas. There is a 38% annual growth rate in Internet from 2016 to 2023 [2, 3], and mobile connectivity [2] is at peak in year 2022. Over 50 million devices around the world will be connected as a part of IOT by 2030.

So, a new solution must be implemented as available frequency resources are limited in 5G demand for bandwidth increases as new applications are growing very fast and the use of mobile data traffic also increases with time. The Internet traffic in 2030 will be 5000 exabytes as per the International Telecommunication Union (ITU) [4]. So, there is a need for revision in the utilization of supplementary frequency bands for higher spectrums. There are lots of challenges

*Current and Future Cellular Systems: Technologies, Applications, and Challenges*, First Edition.
Edited by Garima Chopra, Suhaib Ahmed, and Shalli Rani.

**Table 1.1** Country wise deployment of sub 6GHz frequency bands for 5G Network.

| Countries | 3–4 GHz | 4–5 GHz | 5–7 GHz |
|---|---|---|---|
| China | 3.3–3.6 GHz | 4.5–5 GHz | |
| UK | 3.4–3.8 GHz | | |
| USA | 3.7–3.98 GHz | 4.49–4.99 GHz | 5.9–7.1 GHz |
| Canada | 3.65–4.0 GHz | | 5.9–7.1 GHz |
| India | 3.4–3.6 GHz | | |
| Australia | 3.34–3.7 GHz | | |
| Italy | 3.6–3.8 GHz | | |
| Malaysia | 3.5 GHz | | |
| Korea | 3.4–3.7 GHz, 3.7–4.0 GHz | | |
| Japan | 3.6–4.1GHz | 4.5–4.9 GHz | |
| EU | 3.4–3.8 GHz | | 5.9–6.4 GHz |

that exist in using Millimeter Wave (mmW) bands which hamper its large-scale implementation such as the behavior of mmWs spreading in different directions whenever changes occur in current cellular systems, mmW bands can be affected by multiple objects due to multiple objects, Rapid fluctuations in channels due to small coherence time of mmW bands, etc. The cellular network frequency bands are shown Table 1.1.

### 1.1.1 Motivation

Spectrum sharing facilitates the shared use of the spectrum as compared to a fixed allocation of spectrum where primary users can access the frequency bands with dedicated frequency portions under specific rules and conditions. Spectrum sharing may look trivial but characteristically it is very complex to implement. As Mobile Network Users (MNOs) are operating concurrently, spectrum sharing limits the activities of primary users on the band. So for participation in spectrum sharing schemes and viewing these participations, proper incentives are needed to motivate primary users that impact positive outcomes on the organization.

### 1.1.2 Literature Review

In literature, investigation has been done for spectrum sharing mechanism for CCN and CRSN. Review work for spectrum sharing and cognitive radio is shown in Table 1.2. The table covers description, advantages, and disadvantages of each article.

**Table 1.2** Review works in Spectrum Sharing and Cognitive Radio.

| Ref No. | Author | Description | Advantages | Disadvantages |
|---|---|---|---|---|
| [5] | Ali and Hamounda | Presented a noval spectrum sensing system with proper spectrum sensing methods and its classification. The standard spectrum sensing techniques challenges has also been discussed here | The interweave cognitive radio mode technology for the spectrum sensing for improving the system performance | In the presence of practical imperfections, the compressed sensing based approaches cannot give good results |
| [6] | Sharma, Lagunas et al. | This work provides a review of application of CS in CR communications and compare it with different state of art approaches | Various application areas such as wideband SS and REM construction has been acquired using RF parameter | Less channel occupancy, comprehensive estimation mismatch and realistic signal model evaluation |
| [7] | Kliks et al. | Presented an energy efficient CSS algorithm with news ideas of implementing energy efficient model | Reduction of processing node and energy consumption in each node | Quality of service (QOS) can be easily breach with less reliable spectrum scheme |
| [8] | Tanab and Hamounda | Presented a resource allocation algorithm and approaches for CRNs having certain criteria, techniques and network architecture | Proper resource allocation and its component for cognitive radio networks has been explained | Network mobility, channel models, hybrid users and security of spectrum sensing can be improved |
| [9] | Ahmad, Rehmani et al. | CR sensor networks (CRSNs) based resource allocation schemes designed and classified on optimization standards | Centralized resource allocation in CRSN with different scheme, classes, and categories | Distributed and cluster-based frameworks can also yield good results |
| [10] | Tsiropoulos, Dobre et al. | Review of dynamic spectrum allocation and aggregation with various optimized CRNs based resource allocation designs | Solve the problem of resource allocation using numerous design approaches | Difficult to adapt, reconfigure, accessible the allocation without CRN |

*(Continued)*

**Table 1.2** (Continued)

| Ref No. | Author | Description | Advantages | Disadvantages |
|---|---|---|---|---|
| [11] | Liang, Hanzo et al. | Presented a frequency-division-base and time-division-base channel implement on overlay spectrum access scheme in a cooperative cognitive radio (CCR) network setting | In CCN spectrum scheme both cooperative and non-cooperative games as well as matching games have been shown | Primary users (PUs) and cognitive users (CUs) only interacts with relaying in CCN not with game-theoretical sense |
| [12] | Tehrani, Vahid et al. | Review of characteristics of different authorization regimes with different coordination protocols, network topologies functional on licensed sharing scenarios | Reduces the latency by executing enhanced multi-band scheduling algorithms | Enhanced regulatory regimes and functional requirements is required if MTC services needs to be accommodated in spectrum |
| [13] | Voicu, Petrove et al. | Presented a wireless inter-technology model with a spectrum sharing mechanisms that incorporates a unified, system-level view the technical and non-technical aspects of spectrum | Integration of technical and non-technical aspects at different layers using technology circle | Performance of more than two broadband technologies and deployment of wide networks on inter-technology model without interference is still a big challenge |
| [14] | De Figueiredo, Jiao et al. | Author shows a radio-based framework with optimal use of radio spectrum to achieve improved spectrum utilization | Importance of a pivotal role in spectrum scarcity is explained using time-frequency resources based optimum utilization | Scalable OFDM-based air interface and its slots didn't give fruitful results |

Spectrum sharing facilitated by intellectual radio networks provides a more dynamic and efficient way of spectrum utilization. Intelligent radio enables intelligent autonomous devices to opportunistically sense, optimize, and utilize spectrum availability, improving spectrum efficiency and reducing spectrum scarcity issues Through spectrum sensing is capable of identifying unused or underused spectrum bands, making them more specialized -Reduces interference for users, enabling access to privileged access [15].

Different spectrum sharing models, including underlay, overlay and interweave, address different issues and regulatory requirements. The underlay enables secondary users to broadcast simultaneously with primary users under predetermined interruption limits, while the overlay enables users at the second can opportunistically use spectrum white spaces without harmful interference Primary users rely on interwave spectrum sensing to opportunistically access spectrum when spectrum is not in use [16].

Machine learning (ML) methods play an important role in spectrum sharing capabilities for spectrum sharing, especially in spectrum sensing, decision making, and resource allocation to improve spectrum sensing accuracy, optimize resource allocation, and learning algorithms supervised, unsupervised, reinforcement serves to enhance conscious radio decision-making strategies [17].

Resource allocation in notional radio networks requires optimal allocation of spectrum, power, and other network resources to optimize system performance, ensuring adequacy and efficiency Using method-based optimization, game theory, and ML algorithms are used to improve network efficiency to solve resource allocation challenges [18].

Spectrum sharing in multi-hop/multi-tier millimeter Wave (mmWave) networks is one of the most promising methods that could notably improve capacity and efficiency in such scenarios. Indeed, multiple devices will be allowed to use the same frequency band. Thus, effective scheduling algorithms are indispensable to minimize interference among nodes and provide equal access for all nodes. A scheduling algorithm for multi-hop mmWave networks minimizing transmission time while considering blockage and interference is proposed in [19]. The algorithm efficiently deals with these problems of the factors in mmWave communications. Cross-tier scheduling scheme for multi-tier mmWave wave wireless networks proposed for work in [20] is presented. The proposed scheme aims toward optimization of resource allocation and betterment of overall network performance by coordination of transmissions across different tiers. An optimal link scheduling algorithm for mmWave wave multi-hop networks equipped with multi-user MIMO radios is proposed for work in [21]. The algorithm attempts to maximize network throughput by taking into account the complexities of mmWave propagation and interference.

## 1.2 Spectrum Sharing Technologies

Spectrum-sharing technologies are an important part of for the effective functioning of 5G networks allowing for the effective use of available radio frequency spectrum, to support the anticipated massive increase in data traffic and connections. This technology would help allow for the existence of multiple wireless communications in the same frequency band, maximize management of spectrum efficiency, and help provide and support a wide range of services from high-speed broadband to IoT devices. Here are some of the key spectrums sharing [5] technologies and approaches being utilized and developed for 5G:

a) *Dynamic Spectrum Sharing (DSS)*
   DSS enables 5G services to run in the same frequency ranges as 4G LTE by dynamically distributing spectrum resources in real-time based on demand. This enables operators to deploy 5G networks more quickly and efficiently, using existing spectrum allocations without the need for clearing bands or reassigning frequencies.
b) *Licensed Shared Access (LSA)/Spectrum Access System (SAS)*
   LSA and SAS are frameworks that allow access to underutilized military and government-owned spectrum to commercial users allowing the use of dynamic database and spectrum sensing technologies to temporarily allocate a spectrum to 5G operators. This also helps increase the number of spectrums available without disturbing the incumbent users.
c) *Citizens Broadband Radio Service (CBRS)*
   CBRS uses a 3-tier access system to manage spectrum sharing between incumbent users, giving priority access licenses, having general authorized access for users, and promoting efficient use of spectrum while protecting from harmful interference. It is a specific implementation of SAS in the United States operated in 3.5GHz band.
d) *Millimeter Wave (mmWave) Spectrum Sharing*
   The mmWave frequency is gaining popularity as 5G is expands offering vast amounts of underutilized spectrum. Various techniques can be used for sharing these high frequency bands such as beamforming and directional antennas which can spatially isolate transmission ensuring the sharing of the same frequency band without interfering with other connections.
e) *Cognitive Radio (CR)*
   Cognitive radio is a relatively advanced spectrum that can enhance the spectrum by causing an increase in its efficiency and dynamic interaction. CR can sense their surroundings, locate unoccupied frequency bands, and adjust their transmission parameters in real time to maximize available spectrum without disturbing the incumbent users

f) *Coordinated Multi-Point (CoMP) Operation*
CoMP can be used in crowded network environments for management interference and increasing efficiency. By coordinating signal transmission and reception across multiple base stations, networks can reduce interference while increasing the overall capacity and performance of 5G services.
g) *Non-Orthogonal Multiple Access (NOMA)*
By using varying different power levels and signal processing techniques NOMA enables users to share the same frequency bands improving the spectral efficiency and capacity of 5G networks. It also allows for a greater number of users on the same spectrum band.
Such solutions are only part of a bigger approach to meet 5G requirements which include high data rates, low latency and capacity to serve a large number of devices at the same time. With 5G advancing and new sharing spectrum methodologies and innovations getting created, further improvement will be seen in the efficiency and performance of the wireless networks.

### 1.2.1 Machine Learning in Spectrum Sharing

ML plays an important role in spectrum allocation, enabling decision-making and resource optimization in dynamic and complex wireless environments [22].

ML has been increasingly explored in the context of spectrum sharing, particularly in cognitive radio networks, where intelligent devices dynamically access and utilize available spectrum bands. Here are some existing works in this field along with references:

- Q-Learning-based Spectrum Sharing [23]
- Reinforcement Learning for Spectrum Allocation [24]
- Deep Reinforcement Learning for Spectrum Management [25]
- Game Theory in Spectrum Sharing [26]
- Distributed Learning for Spectrum Sensing [27]
- Fuzzy Logic based Spectrum Allocation [28]

Integrating ML into spectrum sharing can help algorithms predict, learn, and adapt to environments. Here are a few key applications and benefits of ML in spectrum sharing and also shown in Figure 1.1:

a) *Predictive Spectrum Management*
ML can help predict demand and adapt to different spectrum environments ensuring fast spectrum allocation and dynamic sharing. This can help optimize the user experience reducing interference and improving the service for the users.
b) *Cognitive Radio Networks*
Cognitive Radio with the help of ML can help find the most suitable channels and frequencies in the wireless spectrum without disturbing existing users.

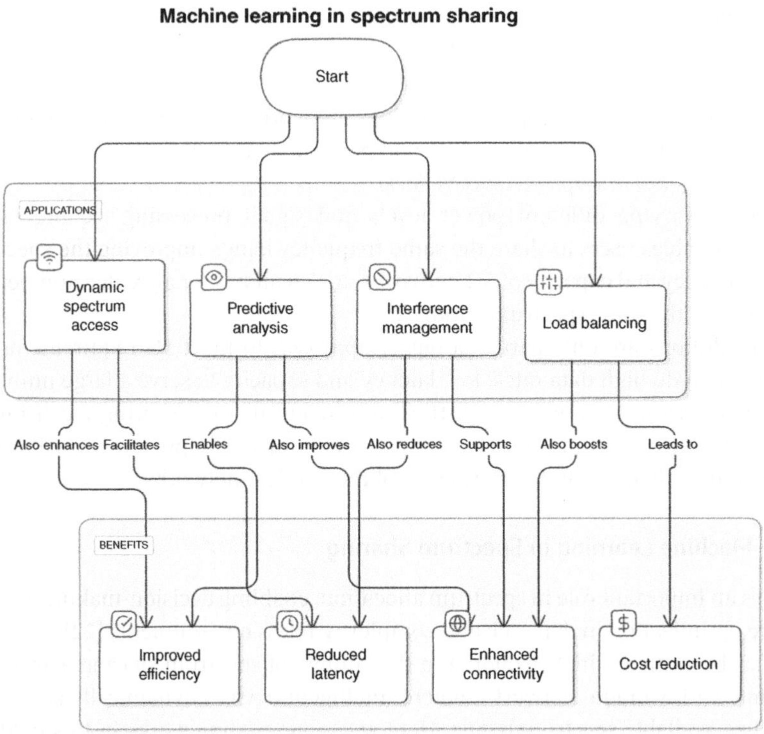


**Figure 1.1** Machine learning in spectrum sharing.

This improves overall performance and plays a key role in the complex process that supports 5G networks.

c) *Interference Management*
ML algorithms can be developed to help identify all potential sources of interference in the frequency band and develop proactive prevention strategies. This is an important feature because it makes the network more reliable by ensuring the smooth operation of various connected devices.

d) *Enhanced Security in Spectrum Sharing*
Security is an important focus as with multiple users on the same frequency range there is a chance that others with malicious intent may misuse the information. ML can be used here to ensure that no interference occurs to detect any irregular patterns and deploy a timely response to prevent the security threat.

e) *Resource Allocation and Optimization*
ML can be used to improve the system for spectrum resource allocation, optimize it for the best performance, and fine-tune it according to the user's behavioral patterns for improved efficiency. It calculates the appropriate timing for spectrum transfer and distributing bandwidth among users.

f) *Signal Classification and Identification*
ML can identify different signals and services received and classify them, distinguishing between the primary and secondary i.e. licensed and unlicenced users respectively. This facilitates proper functioning of the spectrum and prevention of interference in the spectrum signal.

g) *Demand-Responsive Spectrum Access*
The Demand-Responsive spectrum can be enabled across the spectrum with the help of ML adjusting allocation to the spectrum in real-time based on the patterns. This supports efficient and flexible spectrum sharing and plays an important role when spectrum demand fluctuates.

### 1.2.2 Cooperative and Cognitive Radio Networks

Networking and radio intelligence represent two major advances in wireless communications each meeting the growing need for more efficient use of spectrum, improving network performance and increasing wireless connectivity in an age of limited resources [6, 7]. Let's take a closer look at both concepts to understand their capabilities, benefits, and how they complement each other in today's wireless networks.

a) *Cooperative Radio Networks*

Cooperative wireless networks benefit from cooperation from various network nodes by nature in improving the elements of communication reliability, coverage extension, and overall performance in a wireless network. This is an essential principle because it presupposes that the nodes of the network act as relays that assist in transferring information to and from places. Important aspects concerning radio communications include [29, 30]:

1) **Spatial Diversity:** It makes use of varied relay node types that act as space diversities for transmitting the signals to the spectrum. This is effective in reducing fading effects and promoting signal reliability caused by the weak strength of wireless communications.
2) **Expanded Coverage:** Relays can be applied to increase the coverage's scope that a network offers. More so, it remains very useful for emergency situations even in hard-to-reach areas.
3) **Enhanced Spectral Efficiency:** As such, it encourages one communication and hence bigger efficiency with less power needed for the transmission of the signals, making more use of high frequencies feasible.
4) **Enhanced Throughput and Reliability:** By combining signals from multiple paths, cooperative networks can achieve higher throughput and lower error rates compared to non-cooperative approaches.
5) **Cognitive Radio Networks:** Cognitive power and expertise modification of transmission parameters based on temporal spectrum and user requirements

lead to an improvement in the power of access and capacity of spectrum control.

6) **Spectrum Sensing:** Radio information operationalized that the efficient use of spectrum took advantage of access to most times, the frequency bands left unused by primary users.
7) **Spectrum Management:** It recommends the fitting of the most frequency bands for the business and consideration in respect of required specifications as to the availability.
8) **Spectrum Mobility:** Cognitive radios can change their frequency of operation to switch between spectrum bands, ensuring seamless communication even when the current band becomes unavailable or suboptimal.
9) **Spectrum Sharing:** It makes it more efficient and effective in that it's dynamic sharing between users based on a pattern of usage and demand.

#### 1.2.2.1 Integration of Cooperative and Cognitive Radio Networks

Radio technology and telecommunications can be used to improve the performance through direct communication and using dynamic spectrum [8, 31, 32]. This collaboration could lead to:

a) **Boost Spectrum Reach and Efficiency:** While technological collaboration technologies can efficiently analyze spectrum even in a complicated spectrum environment, intelligence is able to detect underutilized spectrum.
b) **Increased Communication Dependability:** It can be attained by integrating technical expertise and coordination, which can lower interference and signal loss and improve communication.
c) **Adaptive Network Topology:** Based on the conditions of the network and the availability of spectrum, the network can modify its topology and communication quality to optimize coverage and performance.

The creation of communications services that satisfy present needs requires the integration of radio networks and radio intelligence for high speed, reliability and efficiency. As technology continues to evolve, the integration of these two approaches will play an important role in the effective management of spectrum and distribution of wireless communications resources.

### 1.2.3 Interference Mitigation Strategies

Interference mitigation gains significant importance in wireless communications due to increasing pressure on the demand for bandwidth and the proliferation of wireless devices in commercial as well as personal space. For instance, effective interference mitigation strategies are important for service delivery, spectral efficiency, and reliable communication in a congested environment [8, 9].

Following are some of the key strategies that help reduce the amount of interference in wireless networks [33, 34]:

1) *Frequency Planning and Management*
    a) **Dynamic Frequency Selection (DFS):** Through automatic selection of relatively less congested frequencies, DFS minimizes interference with other systems.
    b) **Spectrum Allocation:** It includes the frequency range that is being managed in such a way as to ensure there is minimal or no overlay; it has gone to the extent of assigning specific frequencies or groups of bands to different service providers or operators.
2) *Power Control*
    a) **Transmit Power Control:** In this system, the power is controlled in such a manner that one can reduce the level of interference being occasioned on other devices at the same time maintaining an acceptable quality standard of signal.
    b) **Adaptive Power Control:** Dynamically carried out by gaining the maximum link signal strength and reducing its interference in real time based on feedback of effectiveness.
3) *Beamforming and MIMO (Multiple Input Multiple Output)*
    a) **Beamforming:** It focuses patterns of the array of antennae electronic signals towards intended receivers only and makes interferences minimal to and from other unrelated ones.
    b) **MIMO:** Diversifies signals over multiple antennas at the transmitter and the destination. It uses the process of spatial multiplexing for communication that increases performance and decreases interference.
4) *Cognitive Radio and Dynamic Spectrum Access*
    a) **Spectrum Sensing:** There are techniques based on tracking spectrum holes or portions of the spectrum not being used by the licensed users and use these temporarily in such a way that there is no interference to the licensed user.
    b) **Dynamic Spectrum Access:** The system dynamically changes transmission parameters according to the spectrum environment to minimize interference.
5) *Cooperative Communication*
    Relaying nodes used during communication so that they may cooperate to control the interference caused due to the better path of transmission and spatial diversity.
6) *Interference Cancellation Techniques*
    a) **Successive Interference Cancellation (SIC):** A receiver-side technique where the strongest interferers are decoded and subtracted one after the other, leaving at every stage a? weaker desired signal to be detected.

   b) **Joint Signal Processing:** At the receiver, advanced algorithms are used in separating colliding signals from different transmitters.
7) *Inter-Cell Interference Coordination (ICIC) and eICIC for Cellular Networks*
   a) **Inter-Cell Interference Coordination:** A mechanism used today with LTE and by all other cellular networks for controlling interferences among cells, for example, coordinating resource block usage.
   b) **Enhanced ICIC (eICIC):** It involves an added set of strategies like ABS (Almost Blank Subframe). Here the cells are, on some subframes, given the duty of reducing their transmission power so that interference with other users in neighboring cells may be reduced to the minimum.
8) *Network Slicing and Virtualization*
   a) **Network Slicing:** A mechanism of separating various service offerings by flexibly partitioning the physical infrastructure into many virtual network segments with dedicated resources for interference management.
   b) **Virtualization:** The deployment of two recent technologies, network function virtualization (NFV) and software-defined networking (SDN), for optimizing resources that work toward interference.

## 1.3 Case Study and Performance Evaluation

Case studies and performance evaluations play a crucial role in understanding the effectiveness, challenges, and real-world applicability of various technologies and strategies in wireless communication, especially in areas such as spectrum sharing, interference mitigation, and the deployment of cooperative and cognitive radio networks [12]. These studies provide insights into practical implementation issues, performance metrics, and the potential for scalability and adaptation in different environments. Below are some illustrative examples of case studies and performance evaluations in these areas:

1) *Spectrum Sharing in 5G Networks*

   **Case Study:** The performance evaluation in DSS between the military radars, including the incumbent apparatus to the satellite ground stations and the new generation of commercial 5G services, thereby utilizing the licensed spectrum. One such example of technology that uses environmental changes to dynamically change frequencies is called Cognitive Radio [35].

   **Performance Metrics:** They include spectrum utilization efficiency, throughput, latency, and the impact on incumbent users. The performance evaluation presents one clear view of how DSAS can create effective access for a range of different users while protecting incumbents and gaining maximal commercial use of the band.

2) *Interference Mitigation in Dense Urban Environments*
   **Case Study:** Advanced interference management methods are examined through intercell interference coordination (ICIC) and enhanced ICIC (eICIC) [36].
   **Performance Evaluation:** Effectiveness through increased signal-to-interference-plus-noise ratio (SINR), enhanced user throughput, and improved overall network capacity. Results show that coordinated scheduling with power control is very promising in terms of efficient interference management in heterogeneous networks.
3) *Cognitive Radio Networks for Rural Connectivity*
   **Case Study:** Deploy cognitive radio networks to provide broadband connectivity in rural areas by taking advantage of unused TV white spaces. They do so with the help of a best-way study of accessing the underutilized spectrum dynamically and with no interference to the incumbent broadcaster [37].
   **Performance Evaluation:** Based on the coverage area, data rates at the end-user, accuracy of spectrum sensing, and the network's ability to adapt to dynamic changes in the availability of the spectrum. Based on this case study, one can then paint a picture of just how cognitive radio technologies may perhaps end up bridging a certain digital divide concerning the provisioning of such services for remote and underserved areas.
4) *Cooperative Communication for Emergency Response Networks*
   **Case Study:** Cooperative techniques in an emergency response scenario where traditional communication infrastructure does not exist or is compromised. Illustrates deploying a Mobile Ad Hoc Network (MANET) having relay nodes to facilitate communication among first responders [38].
   **Performance Evaluation:** For the evaluation in these simulations, the performance metrics will be network connectivity, end-to-end latency, data throughput, and the robustness of the communication links considered over several states. The results have shown clear benefits for cooperative communication, which can achieve tangible gains in both higher reliability and coverage for mission-critical communication networks.
5) *Network Slicing for IoT Applications*
   **Case Study:** One such new 5G network that includes the network slicing implementation to cover all sorts of IoT application support: from low-power wide-area networks (LPWANs) to high throughput IIoT applications [39].
   **Performance Evaluation:** This work evaluates the performance of IoT in meeting various levels of service-level agreements (SLAs), such as low latency, high throughput, and reliability. The study brings out the efficacy of network slicing in facilitating the accommodation of heterogeneous IoT services.

Together with its attribute metrics and its simulations, they contribute to the validation not only theoretically proposed models but also, to a great extent, empirical and performance studies done in real life. They guide ongoing research efforts and development work on how to meet the requirements of the future, i.e., how wireless communication technologies will evolve.

## 1.4 Future Trends and Challenges

The wireless communication scenario is one replete with a very high level of dynamism and constant shift with the unabated changes in technology accompanied by an ever-expanding need for information, leading therefore to an unbridled desire for universal connectivity [14]. Against this background of change are several new trends and issues shaping the development and deployment of wireless networks. In this respect, it becomes essential to appreciate these trends if one is going to quantify the direction of both research and policies regarding development.

a) **Implementation of AI and ML:** AI and ML are going to be the keystones in the optimization of the network functioning because of predictive maintenance or anomaly detection, dynamic spectrum management, and elimination of interference. However, they can help make real-time decisions from complex patterns and historical data, thus turning them efficient and reliable.
b) **Technological Advancements in Spectrum Sharing:** Although the demand for wireless bandwidth is always rising, the use of available spectral bands becomes almost a compulsion. Advanced technologies like cognitive radio and dynamic spectrum access shall be further developed so that they are made even more flexible, dynamically using the shared spectrum.
c) **Proliferation of IoT and Machine-Type Communications:** The way the IoT is coming up so fast is introducing machine-type communications that will keep growing, hence an exponential increase in connected gadgets. This will make demands on the networks with the capacity to support massive connections of many such devices themselves, each with varying and wide-ranging requirements regarding bandwidths, latencies, and power efficiencies.
d) **The 6G and Beyond Deployment:** World communication gurus have put their effort into researching 6G technology, and it is said that a wider range of data rates, lower delays, and communications that are more trustworthy will be expected. In achieving these massive advancements, some highly potent bleeding-edge technologies are essential for 6G; these include sub-terahertz (THz) frequencies, advanced MIMO, and intelligent reflective surfaces.
e) **Advanced Privacy-Preserving and Security Measures:** The world gets more and more dependent upon wireless communication not only from that critical infrastructural point but also from an individual device angle. As such,

the network security and privacy problems will have to be kept at bay to even higher levels. To this end, any future development has to encapsulate the cryptographic techniques, structures of network security, and privacy-preserving protocols at the cutting edge.

### 1.4.1 Challenges Facing Wireless Communication

a) **Spectrum Scarcity:** Although these new approaches for spectral sharing advance through better technology for spectrum sharing, to date there is no good solution to the inherent problems in the finite natural resource and increasing demand on the wireless services. Hard research and regulatory efforts have to be made to find newer bands up to mmWave and THz frequency for wireless communication.
b) **Energy Efficiency:** Energy consumption by wireless networks is increasingly going to come under concern, especially with the ever-accelerating presence of data centers and network infrastructure. There can be growth in this field only when increasing energy efficiency technologies and protocols are developed.
c) **Interference Management:** Due to the densification of wireless networks, the number of devices coming in contention for the same spectrum increases further. Effective interference management becomes more intricate, and advanced interference mitigation techniques become the cornerstone in maintaining network performance.
d) **Network Security and Privacy:** Wireless networks are so complex and pervasive that several vulnerabilities lurk into them. However, while trying to adapt to these security challenges, a major concern is to ensure performance and privacy of users.
e) **Infrastructure Deployment and Maintenance:** The deployment and maintenance of the required infrastructure of next-generation wireless networks in rural areas or at places far from better accessibility involves logistic and financial issues.
f) **Standardization and Regulation:** Inevitably, such global interoperability and harmonization of the spectrum require constancy in regulatory bodies, industry stakeholders, and standardization organizations. Indeed, such balancing is required for normal development in any country.

Apart from this, there are many such promising opportunities and challenges that lie ahead for the future of wireless communication. Therefore, academia, industry, and world governments will need to join hands in tackling all these. With worldwide collaboration and investment in research and development, the coming era of wireless communication can be prepared to meet the growing needs of humankind.

## 1.5 Conclusion

This chapter thoroughly canvasses the technologies of spectrum sharing, ML applications, cooperative and cognitive radio networks, strategies for interference mitigation, performance evaluation, case studies, future trends, and arising challenges. The advances of these areas define future trends in wireless networks: ways it should be developed and the path along which evolution must be able to proceed, to continue meeting ever more demanding requirements for higher data rate; lower latency; higher reliability; and more and more efficient use of the electromagnetic spectrum.

Cognitive radio networks and spectrum sharing technologies have recently emerged as a couple of dominant solutions to the scarcity in frequency problems, allowing more efficient utilization of available frequency resources. ML will set the standard for revolution in network management and optimization along with AI because it brings power today that was simply not available before to predict network conditions with great levels of accuracy, automate difficult decision processes, and uniformly add intelligence throughout all domains of network performance.

New and advanced technologies in mitigation of interference are being brought in to assure that such increasing density of wireless devices and networks does not compromise the quality of service; these involve various techniques like MIMO, beamforming, and most importantly, interference cancellation through Dynamic Spectrum Access.

Case studies, in addition to performance measurements, have been understood and documented to be a very useful tool in verifying or testing such technologies and strategies in real operational environments. Other than proving practical challenges and successes, it lends a hand to future research and developments by offering great insights into best and needed practices.

With 6G on the horizon, this represents another big game changer in wireless communications. The next generation is likely to bring even faster speeds, lower latencies, and more reliable connections that just might bust open a stupendous new set of application areas across realms ranging from VR/AR to autonomous vehicles to the IoT.

Realization of these advancements is, however, not without challenges that include the following: spectrum scarcity, energy efficiency, network security, and digital divide. Summing up, it can be said that wireless communication has a future, which is very interesting and hard at the same time. It is capable of largely transforming the way one lives and works towards our interaction with the world. The way forward truly raises the stakes for research scholars, industrialists, policymakers, and regulatory bodies to aggressively put their heads together and work out solutions not only for these emerging challenges but also for new opportunities. Undoubtedly, advancement in the wireless network would play a bigger role in rolling out the red carpet for coming decades in social as well as economic developments.

## References


1 Sisinni, E., Saifullah, A., Han, S. et al. (2018). Industrial internet of things: challenges, opportunities, and directions. *IEEE Transactions on Industrial Informatics* 14 (11): 4724–4734.

2 Statista (2022). Percentage of mobile device website traffic worldwide from 1st quarter 2015 to 4th quarter 2021. (Accessed: May 6, 2022). https://www.statista.com/statistics/277125/share-of-website-traffic-comingfrom-mobile-devices/.

3 CISCO (2022). Highlights of Cisco's internet traffic report and forecast. (Accessed: May 6, 2022). https://www.cisco.com/c/dam/m/en_us/solutions/service-provider/vni-forecast-highlights/pdf/Global_2021_Forecast_Highlights.pdf.

4 I. Union (2015). IMT traffic estimates for the years 2020 to 2030. ITU, Geneva, Switzerland, Rep. 2370, Jul. 2015.

5 Ali, A. and Hamouda, W. (2017). Advances on spectrum sensing for cognitive radio networks: Theory and applications. *IEEE Communications Surveys and Tutorials* 19 (2): 1277–1304.

6 Sharma, S.K., Lagunas, E., Chatzinotas, S., and Ottersten, B. (2016). Application of compressive sensing in cognitive radio communications: A survey. *IEEE Communications Surveys and Tutorials* 18 (3): 1838–1860.

7 Cicho'n, K., Kliks, A., and Bogucka, H. (2016). Energy-efficient cooperative spectrum sensing: a survey. *IEEE Communications Surveys and Tutorials* 18 (3): 1861–1886.

8 Tanab, M.E. and Hamouda, W. (2017). Resource allocation for underlay cognitive radio networks: a survey. *IEEE Communications Surveys and Tutorials* 19 (2): 1249–1276.

9 Ahmad, A., Ahmad, S., Rehmani, M.H., and Hassan, N.U. (2015). A survey on radio resource allocation in cognitive radio sensor networks. *IEEE Communications Surveys and Tutorials* 17 (2): 888–917.

10 Tsiropoulos, G.I., Dobre, O.A., Ahmed, M.H., and Baddour, K.E. (2014). Radio resource allocation techniques for efficient spectrum access in cognitive radio networks. *IEEE Communications Surveys and Tutorials* 18 (1): 824–847.

11 Liang, W., Ng, S.X., and Hanzo, L. (2017). Cooperative overlay spectrum access in cognitive radio networks. *IEEE Communications Surveys and Tutorials* 19 (3): 1924–1944.

**12** Tehrani, R.H., Vahid, S., Triantafyllopoulou, D. et al. (2016). Licensed spectrum sharing schemes for mobile operators: a survey and outlook. *IEEE Communications Surveys and Tutorials* 18 (4): 2591–2623.

**13** Voicu, A.M., Simi'c, L., and Petrova, M. (2019). Survey of spectrum sharing for inter-technology coexistence. *IEEE Communications Surveys and Tutorials* 21 (2): 1112–1144.

**14** De Figueiredo, F.A., Jiao, X., Liu, W. et al. (2018). A spectrum sharing framework for intelligent generation wireless networks. *IEEE Access* 6: 60704–60735.

**15** Parvini, M., Zarif, A.H., Nouruzi, A. et al. (2023). Spectrum sharing schemes from 4G to 5G and beyond: protocol flow, regulation, ecosystem, economic. *IEEE Open Journal of the Communications Society* 4: 464–517.

**16** Chang, H.-H., Song, Y., Doan, T.T., and Liu, L. (2023). Federated multi-agent deep reinforcement learning (Fed-MADRL) for dynamic spectrum access. *IEEE Transactions on Wireless Communications* 22 (8): 5337–5348.

**17** Wang, X., Umehira, M., Akimoto, M. et al. (2023). Green spectrum sharing framework in B5G era by exploiting crowdsensing. *IEEE Transactions on Green Communications and Networking* 7 (2): 916–927.

**18** Kim, K.J., Liu, H., Wen, M. et al. (2023). QR decomposition-based cyclic prefixed single-carrier transmissions for cooperative communications: concepts and research landscape. *IEEE Communications Surveys & Tutorials* 25 (1): 133–155.

**19** He, Z., Mao, S., Kompella, S., and Swami, A. (2015). Minimum time length scheduling under blockage and interference in multi-hop mmWave networks. In 2015 IEEE Global Communications Conference (GLOBECOM), December 2015, pp. 1–7. IEEE.

**20** Fan, B., Ramirez, D., Huang, L. et al. (2018). A cross-tier scheduling scheme for multi-tier millimeter wave wireless networks. *IEEE Transactions on Wireless Communications* 17 (8): 5029–5044.

**21** Gómez-Cuba, F. and Zorzi, M. (2019). Optimal link scheduling in millimeter wave multi-hop networks with MU-MIMO radios. *IEEE Transactions on Wireless Communications* 19 (3): 1839–1854.

**22** Ali, S., Saad, W., Rajatheva, N. et al. (2020). 6G white paper on machine learning in wireless communication networks. *arXiv:2004.13875*.

**23** Wang, W., Wang, Y., and Xing, C. (2014). Spectrum sharing for cognitive radio networks based on Q-learning. *Wireless Personal Communications* 79 (3): 1713–1727.

**24** Liang, Y. and Shankar, M. (2017). Reinforcement learning for dynamic spectrum allocation in cognitive radio networks. *IEEE Transactions on Vehicular Technology* 66 (2): 1752–1765.

**25** Mnih, V., Kavukcuoglu, K., Silver, D. et al. (2015). Human-level control through deep reinforcement learning. *Nature* 518 (7540): 529–533.

**26** Cao, Q., Zheng, H., and Guo, W. (2018). A survey on deep learning for radio resource management in wireless networks. *IEEE Communications Surveys & Tutorials* 20 (4): 3137–3168.

**27** Chen, Y., Zhang, W., Zheng, Y., and Lin, C. (2017). Distributed learning for spectrum sensing in cognitive radio networks. *IEEE Transactions on Wireless Communications* 16 (9): 5903–5917.

**28** Zhao, S. and Tian, J. (2017). Fuzzy logic based spectrum allocation for multi-hop cognitive radio networks. *Journal of Network and Computer Applications* 83: 68–75.

**29** Raschellà, A., Eiza, M.H., Mackay, M. et al. (2023). A trust-based cooperative system for efficient wi-fi radio access networks. *IEEE Access* 11: 136136–136149.

**30** A. Kaur and K. Kumar, A reinforcement learning-based green resource allocation for heterogeneous services in cooperative cognitive radio networks. in *IEEE Transactions on Network and Service Management*, vol. 19, no. 2, pp. 1554–1566, June 2022

**31** Singh, S. and Shrestha, R. (2024). A new hardware-efficient and low sensing-time cooperative spectrum-sensor for high-throughput cognitive-radio network. *IEEE Transactions on Circuits and Systems I: Regular Papers* 71 (1): 262–273.

**32** Li, X., Zhou, R., Zhou, T. et al. (2022). Connectivity probability analysis for green cooperative cognitive vehicular networks. *IEEE Transactions on Green Communications and Networking* 6 (3): 1553–1563.

**33** Cho, Y., Yang, W., Oh, D., and Jo, H.-S. (2023). Multi-agent deep reinforcement learning for interference-aware channel allocation in non-terrestrial networks. *IEEE Communications Letters* 27 (3): 936–940.

**34** Skaperas, S., Mamatas, L., and Tsaoussidis, V. (2024). A link-quality anomaly detection framework for software-defined wireless mesh networks. *IEEE Transactions on Machine Learning in Communications and Networking* 2: 495–510.

**35** Biswas, S., Bishnu, A., Khan, F.A., and Ratnarajah, T. (2021). In-band full-duplex dynamic spectrum sharing in beyond 5G networks. *IEEE Communications Magazine* 59 (7): 54–60.

**36** Burhanuddin, L.A.B., Liu, X., Deng, Y. et al. (2023). Inter-cell interference mitigation for cellular-connected UAVs using MOSDS-DQN. *IEEE Transactions on Cognitive Communications and Networking* 9 (6): 1596–1609.

**37** Kryszkiewicz, P., Canfield, C.I., Bhada, S.V., and Wyglinski, A.M. (2022). A systems approach for solving inter-policy gaps in dynamic spectrum access-based wireless rural broadband networks. *IEEE Access* 10: 25165–25174.

**38** Aloqaily, M., Elayan, H., and Guizani, M. (2023). C-healthIER: a cooperative health intelligent emergency response system for C-ITS. *IEEE Transactions on Intelligent Transportation Systems* 24 (2): 2111–2121.

**39** Dawaliby, S., Bradai, A., and Pousset, Y. (2019). Distributed network slicing in large scale IoT based on coalitional multi-game theory. *IEEE Transactions on Network and Service Management* 16 (4): 1567–1580.

# 2

# Synergizing 5G, IoT, and Deep Learning: Pioneering Technological Integration for a Connected Future

*Ankita Sharma and Shalli Rani*

*Chitkara University Institute of Engineering and Technology, Chitkara University, Rajpura, Punjab, India*

## 2.1 Introduction

The frequency of cyberattacks has greatly increased as the Internet of Things (IoT) has become more widely used in accounting, communication, health care, and e-commerce systems. According to experts, it is anticipated that by 2020, there will be fifty billion devices that can connect to wireless networks through Wi-Fi and cellular technology. These devices are expected to send two-thirds of the global Internet traffic [1]. With the progression of technology, safeguarding these systems across wireless networks becomes more and more crucial. Intrusions into banking networks, healthcare systems, and IoT devices possess the capability to disturb vital services and result in substantial yearly financial damages. As a result, there has been a rise in research focused on developing more secure online systems, namely intrusion detection systems (IDS) [2–4].

Considering the widespread use of wireless networks for internet traffic and the ongoing implementation of 5G and IoT technologies, it is extremely likely that these networks possess multiple security flaws that might be targeted by malevolent individuals. In order to mitigate these potentially fatal actions, significant endeavours are being expeditiously undertaken to create wireless intrusion detection systems. Recently, deep learning (DL) approaches have been effectively integrated into NIDS applications. 5G, the fifth generation of mobile networks, has the potential to revolutionise various industries and sectors, offering significant benefits to consumers and enabling wide applications in the industrial sector. However, similar to any technological progress, the implementation of 5G networks raises substantial security problems that must be thoroughly assessed to protect this crucial infrastructure. The telecom business is poised for a significant transformation with the introduction of 5G technology. 5G enables previously unimaginable speeds and communication capabilities.

*Current and Future Cellular Systems: Technologies, Applications, and Challenges*, First Edition.
Edited by Garima Chopra, Suhaib Ahmed, and Shalli Rani.

5G is characterised by its reliability, security, and extensive bandwidth capacity, making it a revolutionary technology in modern civilization. It is useful in driverless vehicles, intelligent infrastructure, hospitals, and AI-enabled houses. The combination of 5G and IoT, two very promising future prospects, is revolutionary in academia, business, and families alike. When 5G networks enhance speed, latency, and cost, billions of networked devices will be able to access a vast new IoT environment. The ultimate integration of these technologies will arise from their mutual development. Furthermore, the goal of the attempt is to broaden the range of their integrated application.

By detecting irregularities and trends in network data, these algorithms have the ability to identify possible dangers [5]. Deep neural networks are trained utilising a diverse range of network data samples, including both regular network operations and abnormal activity that may indicate possible cyber risks [6]. DL can replace delegated feature extraction to create a more secure intrusion detection system or firewall [7]. Contemporary IDS utilise deep learning, which is a form of hierarchical learning, to acquire traffic data representations. This enables them to efficiently identify and prevent intrusions in both wireline and wireless networks.

## 2.2 Security Threats on 5G Network

5G presents a variety of potential attack paths due to its complex design and integration of advanced technologies like edge computing and network slicing [8]. An important concern is the heightened exposure of the network to the outer edges, which happens when data is processed closer to the end user instead of being centralised in secure data centres. The process of decentralisation leads to a broad area that can be allowed for attacks, also allows attackers to take advantage of weak nodes in a network. The segment which is non secure can be vulnerable to different attacks, where the networks integrity will be on stake. The risk and security profiles may vary among different segments [9].

The risk has greatly expanded to broader perspective due to the extensive number of 5G-connected devices most of which are IoT devices lacking sufficient security measures. By serving as network access points, these devices facilitate Distributed Denial of Service (DDoS) attacks that overpower systems. Furthermore, it is important to highlight that 5G networks are highly prone to software vulnerabilities, such as zero-day attacks [10] which allow malicious individuals to exploit unpatched vulnerabilities. Furthermore, MITM attacks have a significant influence on 5G networks. The high-speed data transmission capabilities of 5G may facilitate unauthorised individuals in acquiring sensitive information by intercepting data during its passage across the network. Due to the complex and constantly evolving network structure of 5G there is a higher risk of malevolent individuals taking advantage of weak signalling protocols and Application Program Interface (API) which can lead to unauthorised access and eavesdropping [11].

Moreover, to make the deployment of 5G network in larger perspective the probability of supply chain attacks, when malevolent entities leverage compromised hardware or software provided by manufacturers to infiltrate a network without

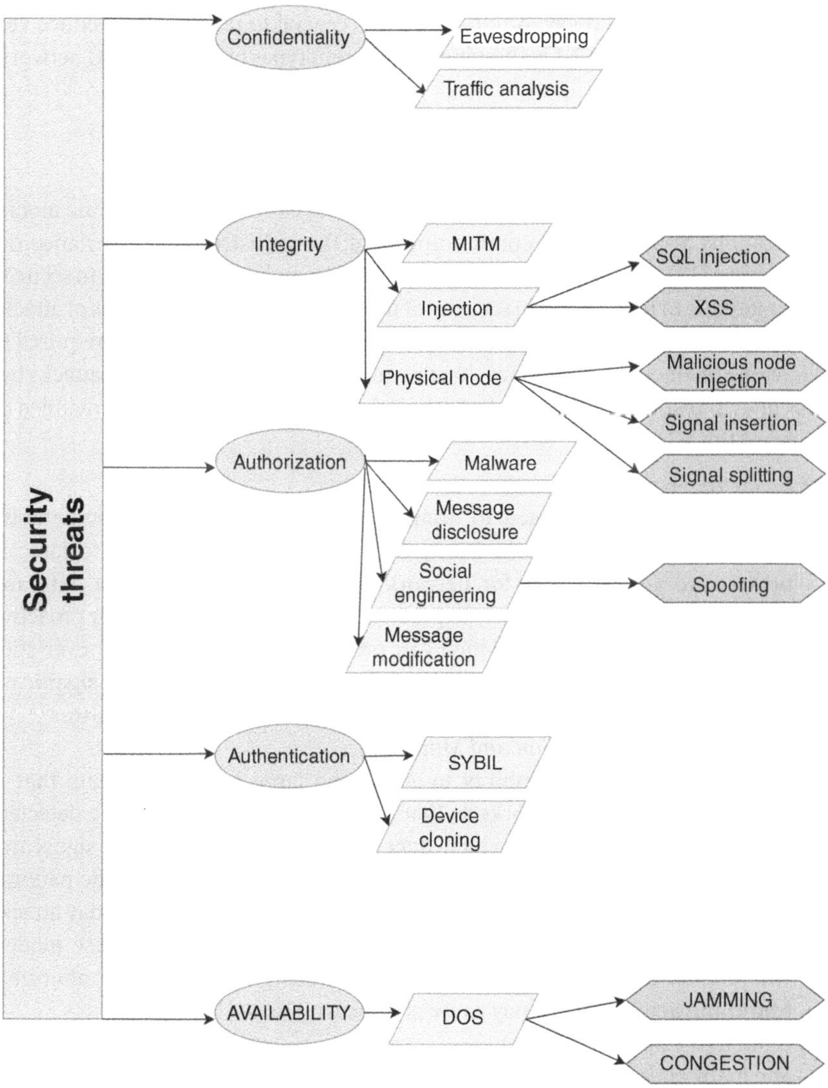


**Figure 2.1** Different types of security threats.

proper authorization. This will make the security of a network weak which will result in the theft of large amount of important data or information [12].

To mitigate these dangers it is crucial that 5G networks incorporate the procedures for security. The encryption for robust networks, also monitoring and detection of ongoing threats having the capacity to breach security at the time of responding to the actions performed. In order to safeguard the future of our extensively networked society, it is imperative to prioritise the security and reliability of 5G networks without delay. Additionally, it is crucial to take steps to reduce vulnerabilities [13]. Figure 2.1 is showing the different types of threats on 5G network.

## 2.3 Applications of 5G

From enhanced connectivity there exist different sectors such as any of the mobile applications also the smart communities and the industrial IoT experience the advantages [14]. The advancements where many complex issues related to security generates due to many flaws in security of networks having strong points of attacks and vary in the structure of the network. The innovative strategies are required to overcome the modern difficulties that not only capable of detecting the attacks but also able to respond to the novel attacks. DL models are highly recommended to improve the 5G security.

1) *Predictive Security Measures*
   DL classifiers can be used to accurately forecast forthcoming vulnerabilities and existing hazards in 5G networks. Algorithms are used to forecast prospective attack targets for network segments by utilising past data and present behaviours. The ability to predict possible dangers allows for proactive strengthening of defences, which can prevent attacks before they ever start and maintain the security of networks [15]. Network slicing is a significant concern that arises from the development of 5G network architectures.
2) *Real-Time Threat Detection and Mitigation*
   DL models have the capability to access the large amounts of data that is transmitted over 5G networks so that anomalies and attacks can be detected. Unlike traditional models which depends upon the pre-existing signatures and set of rules the DL classifiers can identify the harmful and toxic patterns and behaviours. This will strongly help to detect the novel and zero day attacks which were not been detected by the traditional models. Also these models will monitor the traffic of the network continuously also identify the abnormal traffic and activities that may cause any security issue [16].
3) *Enhanced Privacy Prediction*
   For maintaining security and privacy the enormous transmission capacity of data of the 5G networks is mainly responsible. DL classifiers allow the data

packets being marked so that the anomalous packets can be detected and identified thus promoting the use of security and privacy. Moreover, it is feasible to internally train DL models to carry out data anonymization operations within the network. This is accomplished by limiting the quantity of sensitive data that is sent or stored, therefore decreasing the probability of data breaches [17].

## 2.4 Advanced Intrusion Detection Systems (IDS)

DL enhances the ability to identify and categorise data packets that include sensitive information, hence strengthening privacy protections. Compliance with data protection rules may need this. DL models can employ network anonymization of data as part of their learning process. To mitigate the risks of data leakage, one can reduce the sharing or retention of sensitive information [17].

## 2.5 Integration of 5G-IoT-DL

The convergence of the IoT and 5G technology has a significant potential to transform numerous industries. By using the exceptional speed, capacity, and minimal delay of 5G, the security and functionality of IoT may be improved [18]. A crucial element of this integration is the establishment of a resilient infrastructure. In order to meet the requirements of certain IoT applications, especially those that are important and require highly reliable low-latency communication, as well as larger applications that prioritise wide coverage and energy efficiency, it is necessary to build numerous network segments. In addition, edge computing nodes provide instantaneous data analysis and a substantial decrease in latency and bandwidth requirements by handling data from IoT devices in closer proximity to the origin [15]. The operations performed results in this intricate network which depends on a single communication protocol that meets the various requirements of IoT devices in terms of accessibility scalability and energy efficiency [19].

Moreover, the foremost priority for every network is security by utilising advanced and different encryption technologies also required continuous authentication mechanisms. To protect and secure data transfers on the large-scale that occurs with the collaboration of devices in IoT as well as the 5G network [20]. The integration of AI-powered peripheral devices emphasises how crucial it is to make decisions and analyse data quickly in order to take full advantage of 5G's potential and get timely insights [21]. Furthermore, distributed ledger technology improves the integrity and traceability of data by creating a secure ledger system that records data transfers and interactions across the IoT [22]. A special platform that utilises a 5G network will bring about a substantial change in the management of devices

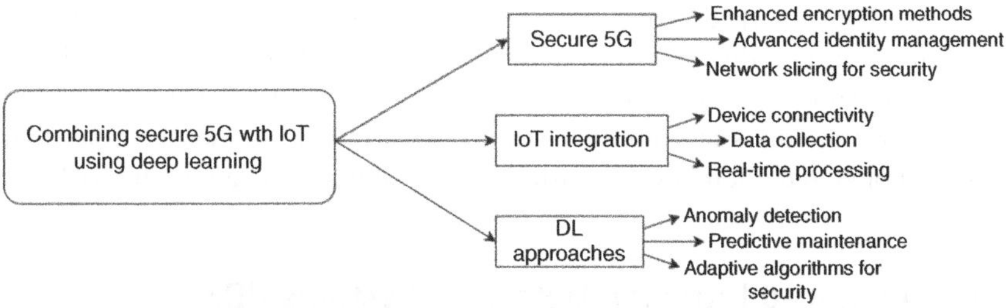


**Figure 2.2** Integration of 5G, IoT, and DL.

allowing for configuration that takes place remotely updating the information and administration of IoT devices [23]. By implementing deep learning-based systems, it will be possible to continuously monitor user activity and network traffic. This will make it easier to quickly discover and resolve any security abnormalities or threats [24]. Figure 2.2 shows the integration of 5G, IoT, and DL

## 2.6 Security Challenges

The integrity of the data stored by an organisation is crucial for the growth of any initiative. When making a decision, it is crucial to take into account the following criteria: data encryption, pathogen detection, and system monitoring. An audit trace is a comprehensive record of all actions executed [25].

Various machine learning (ML) principles, domains, and technologies will enable 5G wireless networks to independently manage their operations, provide services, adjust to higher user and traffic levels, and recover from network failures caused by unexpected events [26]. Figure 2.2 illustrates the expected applications of ML in all aspects of 5G networks. These implementations include the physical and application layers, as well as other services that depend on 5G networks for crucial connectivity [27]. The user's text is enclosed in tags. The emergence of 5G technology brings about new, intricate, and subtle security considerations. Although DL presents distinct difficulties, integrating it into 5G security systems may provide effective solutions [28]. These security issues pertain exclusively to 5G and DL approaches.

1) **Expanded Attack Surface:** 5G networks encompass a wide array of devices, some of which have different levels of security resistance. The diversity of devices and the large number of networked devices in IoT systems greatly increases the potential for attacks [29].
2) **Network Slicing:** 5G technology allows service providers to create several virtual networks within a single 5G physical network. Although it is feasible

to customise each segment to suit the individual requirements of a certain application, ensuring consistent security standards across several segments becomes increasingly difficult, especially in terms of maintaining isolation across segments [30].

3) **Edge Computing:** The connection between 5G networks and edge computing is intimate, since the former evaluates data in greater proximity to its source. Furthermore, other from decreasing data traffic and latency, this approach also disperses data storage and processing, hence presenting new risks [31].
4) **Dynamic Network Configurations:** Given the strict rules like the General Data Protection Regulation (GDPR), DL raises substantial concerns regarding privacy and data protection because it heavily relies on large amounts of data for training and operation [9].
5) **Adaptability to New Threats:** DL algorithms demonstrate exceptional proficiency in recognising existing patterns, although they may face difficulties in efficiently managing zero-day attacks and novel threat vectors that were not part of the training dataset [30]. Cyber threats are constantly changing and developing [32].

## 2.7 Role of ML and DL in 5G at Application and Infra Level

Both ML and DL models are computational systems used to identify distinctive properties of a system that cannot be represented by a mathematical model [33]. The often employed models encompass those designed for categorization, regression, and intelligent agent-environment interactions. After undergoing data training, a model is capable of efficiently performing mathematical computations and making decisions using unfamiliar information [34]. Preserving the present Key will allow for the implementation of 5G data-driven ML models to govern mobility, availability, accessibility, and network communication. The automation of network performance management will lead to enhanced efficiency and effectiveness [35]. ML can be categorised into three main paradigms: (1) supervised learning, where the model is trained by comparing outputs with input samples; (2) unsupervised learning, where the model learns to distinguish input samples even without output labels; and (3) reinforcement learning, where an agent learns to associate any input with an action through interactions with its environment [36]. ML plays a vital role in the operation of wireless networks that enable 5G ML approaches will benefit every level of the network, as shown in Figure 2.3. It is our belief that ML technologies will be used to develop every component of traditional wireless communication systems [37]. Wireless researchers should

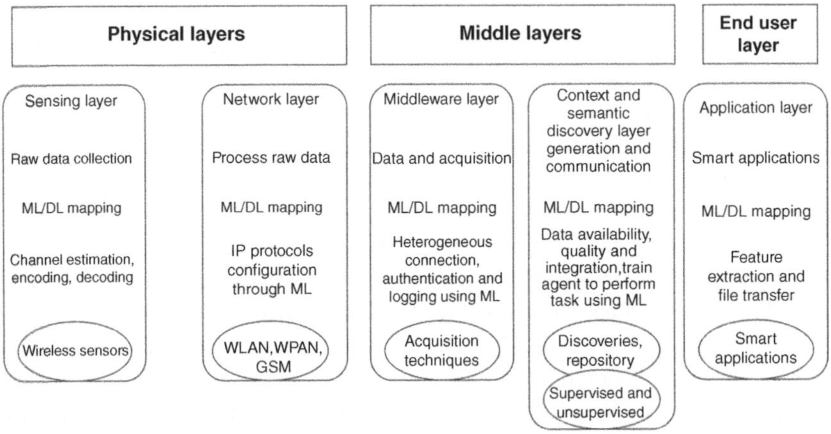


**Figure 2.3** Conceptual working of 5G-DL.

begin exploring potential methods for integrating ML into wireless networks. The provision of training data is crucial for a ML model to achieve the desired results.

At the application level, ML utilises advanced capabilities like facial recognition and user location tracking to relay sensitive information, transforming the way low latency and quicker processing are achieved [19]. ML is revolutionising the provision of low latency and faster processing in response to increasing demand at the application level. This is achieved through the employment of powerful features such as facial recognition and the usage of user location to convey sensitive information [38]. The application layer has three primary objectives: enhancing security and safety through continuous and comprehensive monitoring, and optimising resource management. The software rapidly installs on mobile devices and utilises actuators to modify the living environment of the users in order to meet the specified performance standards [39–42]. Figure 2.4 shows the ML/DL

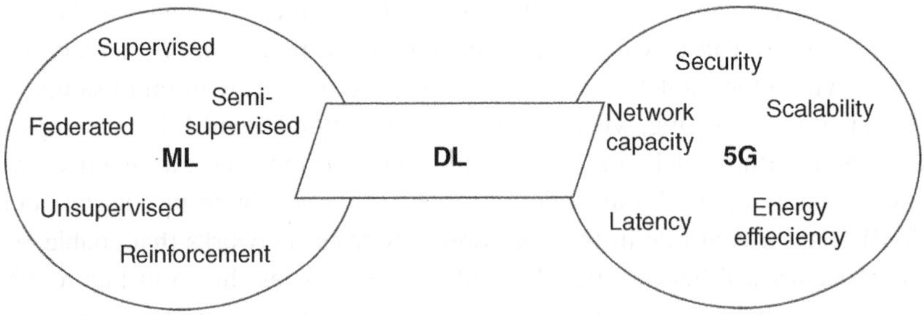


**Figure 2.4** ML/DL at with 5G.

**Table 2.1** Overview of 5G with requirements at user and infrastructure level.

| Factors | Application level | Infrastructure level | Requirements |
|---|---|---|---|
| Data breaches | 1 | 0 | At central locations more chances to breach |
| Latency | 1 | 0 | Data should be at higher speed |
| More privacy and security | 1 | 0 | Using ML/DL more security at user end is needed |
| scalability | 1 | 0 | Rather than 1 central location more storage is required |
| Dependability on devices | 0 | 1 | Mobile apps size should be reduced |

Where 1 = Yes and 0 = No

with 5G. These models involve the distinct modelling and optimisation of various crucial modules. Table 2.1 is showing the overview of 5G with requirements at user and infrastructure level.

## 2.8 Conclusion

In this chapter, we analysed the mechanisms of various ML algorithms. In addition, we examined the benefits and potential future prospects of 5G communication technology. Examining the current advancements in technology, the demand for 5G is compared and evaluated. Research has concluded that the application level is more effective in addressing the challenges arising from 5G problems compared to the infrastructure level. The integration of many ML technologies with a 5G wireless communication network enables efficient and sophisticated functioning. We must develop a comprehensive solution at both the application and infrastructure levels to handle many challenges related to latency, power allocation, privacy, security, and model compatibility in the context of current ML and 5G technology.

## References

**1** Kibria, M.G., Nguyen, K., Villardi, G.P. et al. (2018). Next generation new radio small cell enhancement: architectural options, functionality and performance aspects. *IEEE Wireless Communications* 25 (4): 1–9.

**2** Ahmad, I., Shahabuddin, S., Kumar, T. et al. (2019). Security for 5G and beyond. *IEEE Communications Surveys & Tutorials* 21 (4): 3682–3722.

**3** Al-Garadi, M.A., Mohamed, A., Al-Ali, A. et al. (2020). A survey of machine and deep learning methods for internet of things (iot) security. *IEEE Communications Surveys & Tutorials* 22 (3): 1646–1685.

**4** Arfaoui, G., Bisson, P., Blom, R. et al. (2018). A security architecture for 5g networks. *IEEE Access* 6: 22466–22479.

**5** Liu, Q., Li, P., Zhao, W. et al. (2018). A survey on security threats and defensive techniques of machine learning: a data driven view. *IEEE Access* 6: 12103–12117.

**6** Miller, D.J., Xiang, Z., and Kesidis, G. (2020). Adversarial learning targeting deep neural network classification: a comprehensive review of defenses against attacks. *Proceedings of the IEEE* 108 (3): 402–433.

**7** Haider, N., Baig, M.Z., and Imran, M. (2020). Artificial intelligence and machine learning in 5g network security: opportunities, advantages, and future research trends. *arXiv preprint arXiv:2007.04490* .

**8** Papernot, N., McDaniel, P., Sinha, A., and Wellman, M. P. (2018). SoK: security and privacy in machine learning. in Proc. IEEE Eur. Symp. Secur. Privacy (EuroS&P), London, UK (24–26 April 2018), pp. 399–414. IEEE.

**9** Zhang, C., Patras, P., and Haddadi, H. (2019). Deep learning in mobile and wireless networking: a survey. *IEEE Communications Surveys & Tutorials* 21 (3): 2224–2287.

**10** Yuan, X., He, P., Zhu, Q., and Li, X. (2019). Adversarial examples: attacks and defenses for deep learning. *IEEE Transactions on Neural Networks and Learning Systems* 30 (9): 2805–2824.

**11** Qayyum, A., Usama, M., Qadir, J., and Al-Fuqaha, A. (2020). Securing connected & autonomous vehicles: challenges posed by adversarial machine learning and the way forward. *IEEE Communications Surveys & Tutorials* 22 (2): 998–1026.

**12** Benzaid, C. and Taleb, T. (2020). ZSM security: threat surface and best practices. *IEEE Network* 34 (3): 124–133.

**13** Avizienis, A., Laprie, J.-C., Randell, B., and Landwehr, C. (2004). Basic concepts and taxonomy of dependable and secure computing. *IEEE Transactions on Dependable and Secure Computing* 1 (1): 11–33.

**14** Repko, A.F. and Szostak, R. (2020). *Interdisciplinary Research: Process and Theory*, 4e. Thousand Oaks, CA: Sage.

**15** Ramesh, T.R., Lilhore, U.K., Poongodi, M. et al. (2022). Predictive analysis of heart diseases with machine learning approaches. *Malaysian Journal of Computer Science* 132–148.

**16** Buda, T. S., Assem, H., Xu, L. et al. (2016). Can machine learning aid in delivering new use cases and scenarios in 5G?

**17** Xiong, Z., Zhang, Y., Niyato, D. et al. (2019). Deep reinforcement learning for mobile 5G and beyond: fundamentals, applications, and challenges. *IEEE Vehicular Technology Magazine* 14 (2): 44–52.

**18** Wang, J., Jiang, C., Zhang, H. et al. (2020). Thirty years of machine learning: The road to Pareto-optimal wireless networks. *IEEE Communications Surveys & Tutorials* 22 (3): 1472–1514.

**19** Badotra, S. and Panda, S.N. (2021). SNORT based early DDoS detection system using Opendaylight and open networking operating system in software defined networking. *Cluster Computing* 24 (1): 501–513.

**20** You, X., Zhang, C., Tan, X. et al. (2019). AI for 5g: research directions and paradigms. *Science China Information Sciences* 62 (2): 21301.

**21** Yao, M., Sohul, M., Marojevic, V., and Reed, J.H. (2019). Artificial intelligence defined 5g radio access networks. *IEEE Communications Magazine* 57 (3): 14–20.

**22** Khan, R., Kumar, P., Jayakody, D.N.K., and Liyanage, M. (2019). A survey on security and privacy of 5g technologies: potential solutions, recent advancements, and future directions. *IEEE Communications Surveys & Tutorials* 22 (1): 196–248.

**23** 3GPP (2019). Security architecture and procedures for 5g system. Technical Specifications, 3rd Generation Partnership Project, June 2019.

**24** Lee, J.-H. and Kim, H. (2017). Security and privacy challenges in the internet of things [security and privacy matters]. *IEEE Consumer Electronics Magazine* 6 (3): 134–136.

**25** Riazi, M. S., Weinert, C., Tkachenko, O. et al. (2018). Chameleon: a hybrid secure computation framework for machine learning applications. in Proceedings of the 2018 on Asia Conference on Computer and Communications Security, ASIACCS '18, New York, pp. 707–721, ACM.

**26** Tanuwidjaja, H. C., Choi, R., and Kim, K. (2019). A survey on deep learning techniques for privacy-preserving. in International Conference on Machine Learning for Cyber Security, Xi'an, China (19–21 September 2019), pp. 29–46. Springer.

**27** Bhuyan, M.H., Bhattacharyya, D.K., and Kalita, J.K. (2013). Network anomaly detection: methods, systems and tools. *IEEE Communications Surveys & Tutorials* 16 (1): 303–336.

**28** Hassan, M.U., Rehmani, M.H., and Chen, J. (2019). Differential privacy techniques for cyber physical systems: a survey. *IEEE Communications Surveys & Tutorials* 1–1.

**29** Dai, Y., Xu, D., Maharjan, S. et al. (2019). Blockchain and deep reinforcement learning empowered intelligent 5g beyond. *IEEE Network* 33 (3): 10–17.

**30** Luong, N.C., Hoang, D.T., Gong, S. et al. (2019). Applications of deep reinforcement learning in communications and networking: a survey. *IEEE Communications Surveys & Tutorials* 21 (4): 3133–3174.

**31** Ferdowsi, A., Challita, U., Saad, W., and Mandayam, N. B. (2018). Robust deep reinforcement learning for security and safety in autonomous vehicle systems. in 2018 21st International Conference on Intelligent Transportation Systems (ITSC), Maui, HI (04–07 November 2018), pp. 307–312. IEEE.

**32** Huang, L., Joseph, A. D., Nelson, B. et al. (2011) Adversarial machine learning. in Proceedings of the 4th ACM Workshop on Security and Artificial Intelligence, New York (21 October 2011), pp. 43–58. ACM.

**33** Kalkan, K., Altay, L., Gür, G., and Alagöz, F. (2018). JESS: joint entropy-based DDoS defense scheme in SDN. *IEEE Journal on Selected Areas in Communications* 36 (10): 2358–2372.

**34** Virupakshar, K.B., Asundi, M., Channal, K. et al. (2020). Distributed Denial of Service (DDoS) Attacks Detection System for OpenStack-based Private Cloud. *Procedia Computer Science* 167 (2019): 2297–2307.

**35** Bhushan, K. and Gupta, B.B. (2019). Distributed denial of service (DDoS) attack mitigation in a software-defined network (SDN)-based cloud computing environment. *Journal of Ambient Intelligence and Humanized Computing* 10 (5): 1985–1997.

**36** Huang, K., Yang, L., Yang, X. et al. (2020). A low-cost distributed denial-of-service attack architecture. *IEEE Access* 8: 42111–42119. https://doi.org/10.1109/ACCESS.2020.2977112.

**37** Dong, S., Abbas, K., and Jain, R. (2019). A survey on distributed denial of service (DDoS) attacks in SDN and cloud computing environments. *IEEE Access* 7: 80813–80828.

**38** Lilhore, U.K., Poongodi, M., Kaur, A. et al. (2022). Hybrid model for detection of cervical cancer using causal analysis and machine learning techniques. *Computational and Mathematical Methods in Medicine* 2022: 4688327.

**39** Hayat, O., Ngah, R., Kaleem, Z. et al. (2020). A survey on security and privacy challenges in device discovery for nextgeneration systems. *IEEE Access* 8: 84584–84603.

**40** Morocho-Cayamcela, M.E., Lee, H., and Lim, W. (2019). Machine learning for 5G/B5G mobile and wireless communications: Potential, limitations, and future directions. *IEEE Access* 7: 137184–137206.

**41** Sultana, N., Chilamkurti, N., Peng, W., and Alhadad, R. (2019). Survey on SDN based network intrusion detection system using machine learning approaches. *Peer-to-Peer Networking and Applications* 12 (2): 493–501.

**42** Checkland, P. (1999). *Systems Thinking, Systems Practice*. Chichester, U.K.: Wiley.

# 3

# Driving Next Generation IoT with 5G and Beyond

*Shishir Shrivastava, Ankita Rana, and Ashu Taneja*

*Chitkara University Institute of Engineering and Technology, Chitkara University, Punjab, India*

## 3.1 Introduction

As the technological landscape continues to evolve, the demand for ever-increasing data transfer speeds and network reliability becomes paramount. In this context, fifth generation (5G) technology emerges as a transformative force, promising a significant leap forward in connectivity. Ponder a world where information moves at the wink of an eye. The ability to download movies in seconds and stream virtual reality experiences without buffering are some of the promises that 5G offers. 5G is more than just faster internet it is a total reboot aimed at supporting an Internet-of-Things (IoT) filled globe. Fourth generation (4G) can be compared to a four-lane motorway, it works well in terms of getting all the traffic moving however, with increasing number of smartphones, tablets, smart home appliances and emerging technologies congestion is inevitable. On top of utilizing clever technologies such as massive multiple input multiple output (MIMO), this expressway evolved from a multi-lane highway by expanding it into several lanes. This entails the installation of numerous antennas on both cell towers and mobile devices, which collaborate to transmit and receive a greater amount of data concurrently, much to the addition of extra lanes to facilitate smoother traffic flow.

However, speed alone is not sufficient. Low latency is another impressive feature that 5G has. This is necessary for applications requiring real-time responses. Consider the scenario when there is a negligible time lag between the actions performed by surgeons working from a distance or self-driving cars swiftly adjusting to their surroundings. In addition to reduced bandwidth requirements, this communication pathway seems almost instantaneous, the equivalent of

*Current and Future Cellular Systems: Technologies, Applications, and Challenges*, First Edition.
Edited by Garima Chopra, Suhaib Ahmed, and Shalli Rani.

having a priority lane for emergency messages. For example, the challenge of connecting a huge number of devices has been overcome using 5G. This is where small cells come in. These tiny mobile towers scattered all over the city behave like signal boosters to guarantee consistent connectivity even in crowded areas. It is like adding more exits and lanes to our data highway only that instead of cars it is a network of devices that we manage and they are all seamlessly connected.

5G technology has numerous features which includes faster response time, higher data rates. High capacity, high speeds, capability for easier upgradation and easier integration with other technologies [1]. This makes it suitable for wide range of high connectivity and low latency applications. The features of 5G technology are shown in Figure 3.1.

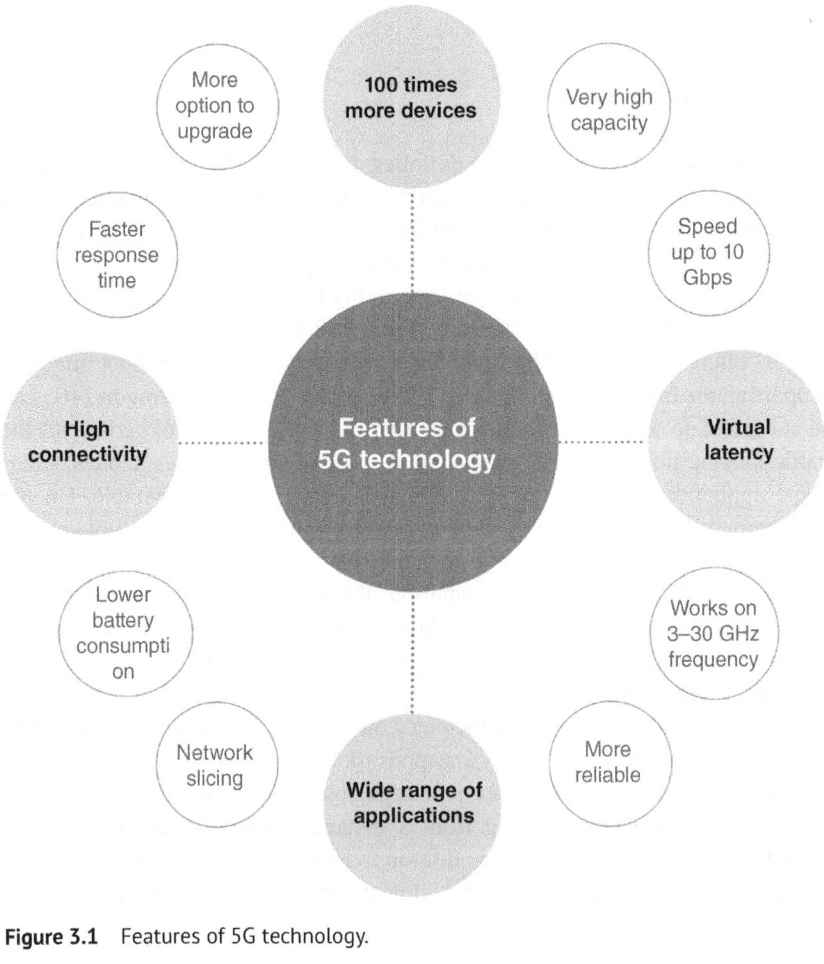


**Figure 3.1** Features of 5G technology.

## 3.2 Need for Technological Advancement

When considering the future scale of IoT, the present 4G network has been efficient in supporting our mobile internet requirements but is not sufficient to fulfil a large-scale IoT future. In this chapter, the development of 5G technology for realizing the full potential of IoT is discussed.

The main cause of the capacity crunch is that 4G can't handle the huge number of devices that IoT will involve. To guarantee that all of the devices are connected at the same time and to keep the network from getting overloaded with devices, 5G is essential. To track and manage all these endpoints would be an unachievable task otherwise. Talking about IoT applications, speed is always important because many of them need for instantaneous, real-time data interchange. Regretfully, 4G latency is insufficient for applications like driverless cars and remote surgery, which need for instantaneous connectivity because even milliseconds might mean the difference between life and death.

Regarding the IoT, battery life is the key problem, mostly because of the resource-constrained sensors. Another crucial issue is that 4G communication methods are thought to be power-intensive. Improves in 5G are necessary to develop efficient communication techniques that prolong the battery life of IoT devices [2].

As IoT expands to encompass billions of interconnected devices, the limitations of 4G networks become evident. 4G simply can't handle the sheer volume of data (data rate) or the constant movement of these devices (mobility). 5G tackles these challenges head-on, offering a significant boost in data transfer speeds to accommodate the data deluge and improved network performance for devices on the move. Security is paramount in this vast network, and 5G incorporates robust measures like encryption and authentication to safeguard data privacy and communication integrity. Additionally, 5G boasts enhanced reliability, minimizing disruptions and ensuring critical data from sensors and devices reaches its destination. Finally, 5G offers improved position accuracy, crucial for applications like asset tracking and emergency response. In essence, 5G isn't just about faster internet; it's a comprehensive upgrade that unlocks the full potential of the IoT revolution, laying the groundwork for a future where billions of interconnected devices work together seamlessly in a secure and reliable environment. Figure 3.2 shows some of the important aspects of 5G.

## 3.3 Existing Wireless Technologies

**Wi-Fi (Wireless Fidelity):** The entire world is cognizant of and enamoured with Wi-Fi. It offers rapid internet connectivity within a restricted local area network

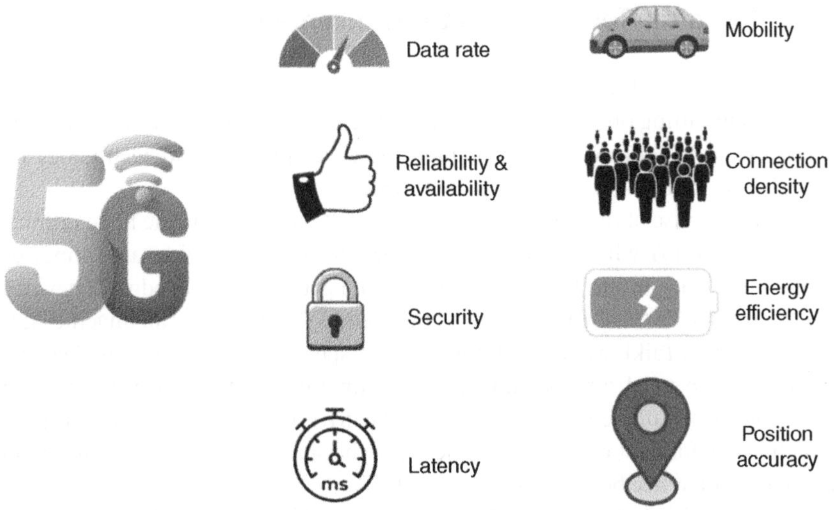


**Figure 3.2** Aspects of 5G.

(LAN) and it is perfect to connect things such as laptops, smartphones, and tablets to the internet. However, Wi-Fi's reach is limited. The Signal strength weakens significantly with distance, and it typically requires a direct connection to a router or hotspot, hindering its use in expansive areas or those lacking infrastructure. Wi-Fi eliminates cables, enabling mobility and easy network expansion. It boasts user-friendly setup, supports multiple devices, and offers robust security protocols.

**LoRa (Long Range)**: This technology mainly focuses on long-range low-power remote sensing for short data packets that travel long distances. Perfectly suited for use in remote areas like agriculture or industry where battery life is critical because these sensors are utilized. While data transfer rates are slower compared to Wi-Fi, long range (LoRa) thrives in situations with poor existing infrastructure or where data transmission occurs at intervals. However, LoRa's limited data packet size restricts the complexity of the information it can carry.

**LTE (Long-Term Evolution)**: It is the current dominant technology for mobile internet access on smartphones and tablets. Long-term evolution (LTE) has satisfactory speed and coverage to browse the web, stream videos, make calls, etc. However, LTE's power consumption can be a drawback. This limitation makes it less ideal for low-power devices like wearables that need extended battery life, pushing advancements towards 5G's promise of greater efficiency alongside even faster speeds.

**LoRaWAN (Long Range Wide Area Network)**: Long range wide area network (LoRaWAN) functions as the main network component for LoRa

**Table 3.1** Comparison of different existing wireless technologies.

| Technology | Wi-Fi | LoRa | LoRaWAN | LTE |
|---|---|---|---|---|
| Data rate | High (up to Gbps) | Low (0.3 kbps - 50 kbps) | Low (0.3–50 kbps) | Medium (10 Mbps–1 Gbps) |
| Coverage | Short (indoor/limited outdoor) | Long (rural: 15 km, urban: 2–5 km) | Long (rural: 15 km, urban: 2–5 km) | Wide (global) |
| Bandwidth | High | Low | Low | Medium to high |
| Range | Up to 100 m | Up to 15 km (rural) | Up to 15 km (rural) | Up to several km |
| Cost | Low to Medium | Low | Low | Medium |
| Security | Can be vulnerable | More secure | More secure | Secure |
| Latency | Low | High | High | Medium to low |
| Applications | Streaming, web browsing | Asset tracking, smart metering | Smart cities, industrial IoT | Wearables, connected cars |

communication. Thus, it is a networking protocol that operates on top of the LoRa technology to ensure safe communicating and network management for a vast number of devices based on LoRa. Therefore, large scale IoT deployments will benefit from this technology because it allows smooth data movement between devices and remote servers. While powerful for wide-area applications, LoRaWAN may not be the best choice for scenarios requiring real-time data exchange due to its slower data transfer rates [3].

The comparison of different existing wireless technologies is shown in Table 3.1.

## 3.4 Challenges in Existing Technologies

As much as they may be of immense assistance, our today's wireless technologies are not without flaws. The more we depend on our gadgets and the internet, the more some limitations start revealing themselves. Some current challenges faced by wireless technologies.

**Short Range:** Technology like WiFi is great for connecting devices within a house or office but its range is quite limited. This can be frustrating if you have a large property or live in rural areas with spotty WiFi coverage. Assume that you are a farmer who needs to check his scattered farms using sensors or you are a park ranger looking to connect devices across an extensive national park. In such cases, existing technologies might find it hard to provide stable connection due to. For instance, expensive extenders or boosters might be necessary to create usable signal while the signal might also remain unreliable along the fringe at this point. Limited area inhibits full development of applications that rely on wider connections.

**Data Bottlenecks:** Networks can get overloaded with more and more gadgets that access the internet by means of streaming high definition (HD) videos or downloading large files. Just think about a highway during rush hour when it is jam packed with cars beyond its capacity and therefore creating congestion, slowdowns, stop-and-go traffic. This tends to be a weak point for activities that require fast and stable internet like video conferencing or online games. When your call gets dropped in the middle of an important presentation or there is lagging during a competitive online game picture, the feeling of frustration arises within you. The technology in place must be capable of accommodating growing needs for data so as to ensure an uninterrupted experience while on line.

**Battery Drain:** For many radio frequency technologies, especially those that offer such high data rates as LTE, battery drain is a major issue. This suggests that if you are constantly online browsing, streaming music, and using GPS, your smartphone battery may run out really quickly. Imagine yourself using your phone to get directions in a new place and discovering it has died from constant data use. In such scenario, you might not be able to securely return to your accommodation or record some significant moments with pictures. More energy efficiency in the current technology is required to extend the battery life on a single charge. While battery technology has to be improved, wireless protocol advancements are equally important to maximize data transfer and reduce power usage.

**Security Concerns:** When we connect more and more personal information and devices to the internet, security becomes a topmost concern. Hackers can take advantage of vulnerabilities in present wireless technologies to seize data or disrupt communication. Your belongings are still at stake when you have a broken door lock. This could be analogous to having a lock on your door but one that is easy to break into; one can still steal from you in such case.

Weak security in wireless networks provide access to your information for hackers who can steal passwords, credit card details or even private messages. Our privacy and data in this wireless world can only be safeguarded through tight security measures. Developments in encryption protocols as well as network authentication techniques are important for ensuring safety when carrying out our online activities.

**Limited Scalability:** Imagine a scenario of smart city where thousands of sensors are installed in order to monitor traffic flow, environmental conditions or even parking availability. Existing technologies might not cope well with such huge number of devices communicating at the same time. A concert venue packed with tens of thousands of people trying to logon onto Wi-Fi system may cause it shut down because its capacity is exceeded by demand. We need wireless systems that can efficiently manage a growing number of connected devices as the IoT continues to expand. Advancements in network architecture and resource allocation techniques are necessary to ensure seamless connectivity even in scenarios with a massive number of devices vying for a connection. These limitations highlight the need for advancements in wireless technologies to keep pace with our ever-growing reliance on connectivity and the vast potential of IoT.

## 3.5 Towards 5G Communication

All people are starting to value mobile internet more and more. Every day life depends heavily on our wireless networks. They enable us to control smart house appliances and stream HD movies. These needs keep expanding, thus 4G technology is becoming outdated. As these demands continue to grow, 4G technology is reaching its limits. This is where 5G, the fifth generation of mobile network technology, steps in.

- **Exponential Data Growth:** Cell phone data transmission is soaring. High-bandwidth applications like virtual reality and augmented reality are made feasible by 5G's far faster data rates (up to 20 Gbps) than 4G's (up to 1 Gbps).
- **Ultra-Reliable Low-Latency Communication (uRLLC):** Response times in applications like self-driving cars, remote surgery, and industrial control must be lightning fast. By cutting the time it takes to transmit data, 5G's extremely low latency (less than 1 millisecond) enables various applications.

To achieve these advancements, 5G utilizes several key technologies listed below in Figure 3.3

### 3.5.1 MIMO and Massive MIMO

MIMO technology serves as a cornerstone for 5G networks, significantly enhancing data transmission. MIMO utilizes multiple antennas on both the base station and user equipment (UE) to transmit and receive data streams simultaneously, effectively multiplying network capacity as shown in Figure 3.4. Imagine a traditional cellular network transformed into a high-speed data highway for significantly faster data transfer. MIMO's core functionality lies in exploiting

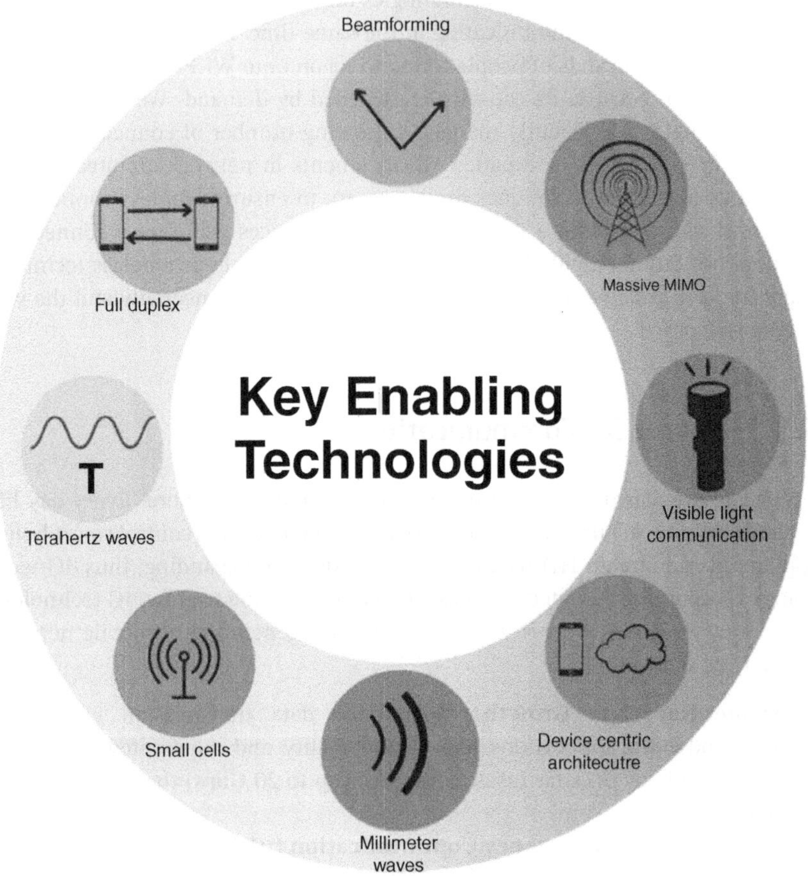


**Figure 3.3** Key enabling technologies.

spatial diversity. By utilizing multiple antennas separated by a distance, MIMO leverages the fact that signal fading and interference can affect each antenna differently. If a signal weakens on one antenna, it might be stronger on another, allowing for robust data transmission. Additionally, MIMO employs techniques like spatial multiplexing and beamforming for further enhancement. Spatial multiplexing transmits multiple data streams simultaneously using different antennas, while beamforming focuses the transmitted signal towards the specific location of the UE, improving signal strength and reducing interference.

A variant of MIMO known as massive MIMO takes these enhancements even further. While both MIMO and massive MIMO utilize multiple antennas, a key distinction lies in the number employed. Traditional MIMO utilizes a moderate

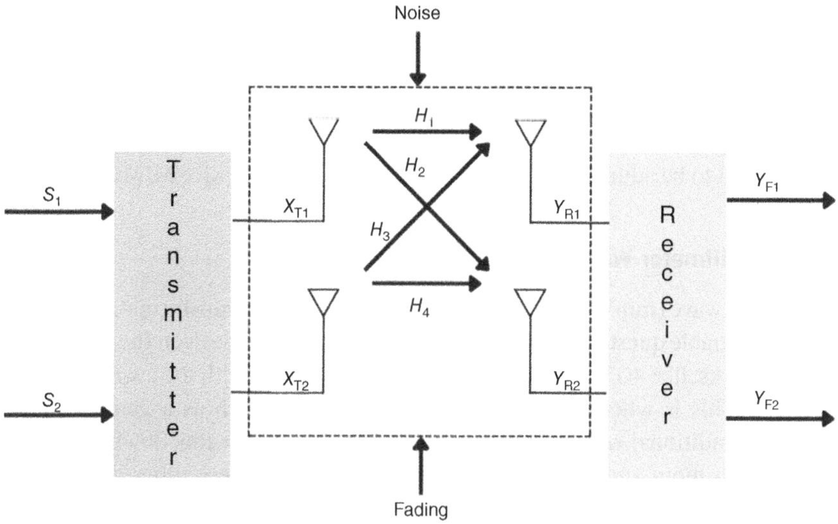


**Figure 3.4** MIMO block diagram.

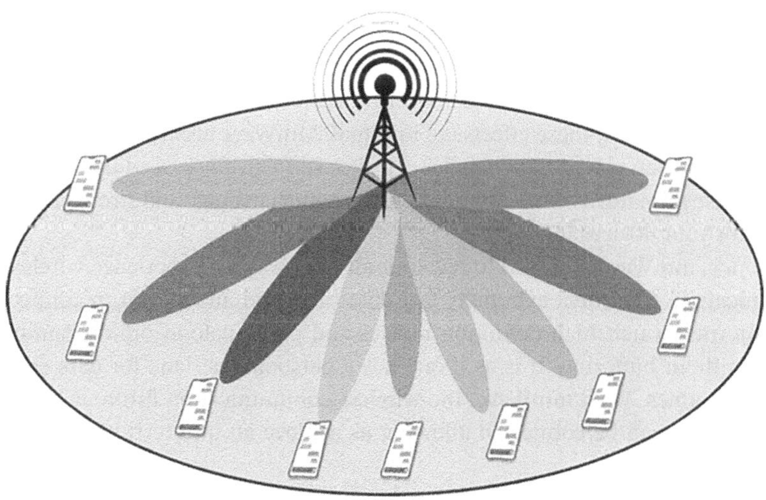

**Figure 3.5** System model depicting massive MIMO.

number, offering substantial improvements. Massive MIMO, on the other hand, deploys a vast antenna array at the base station compared to the UE's smaller number as depicted in Figure 3.5. This dramatic difference unlocks even greater potential. Massive MIMO offers superior performance due to its extensive antenna array. With more antennas spread over a wider area, it offers exceptional spatial diversity, leading to robust and reliable data transmission. Additionally, massive MIMO enables highly sophisticated beamforming techniques, creating focused signal paths directed towards individual UEs. This significantly improves network capacity and user experience by efficiently targeting each device and minimizing interference. However, challenges related to physical space, cost, and processing power need to be addressed for widespread adoption of massive MIMO [4, 5].

### 3.5.2 Millimeter Wave (mmWave) Communication

Millimeter wave (mmWave) communication emerges as a transformative solution in our insatiable quest for data amidst advancing technology. Even the latest wireless networks, like 4G LTE, eventually struggle to keep up with the ever-increasing demand. This is where mmWave communication steps in as a game-changer. Think of traditional wireless networks as a web of data streams flowing through tubes. With more and more devices trying to connect, these tubes get clogged, leading to slow speeds and frustrating buffering. MmWave offers a revolutionary solution, it utilizes much higher frequencies than what is currently used. Imagine these frequencies as much wider tubes, allowing significantly more data to travel simultaneously. This translates to super-fast data speeds, enabling you to download movies in seconds or experience seamless virtual reality without any lag.

There have been concerns about mmWave being susceptible to interference from rain or fog. However, extensive research has revealed that for shorter distances, like within a city block, these effects are minimal. MmWave also thrives in urban environments when paired with smaller cell sites strategically placed throughout the city. These miniaturized base stations ensure better coverage and stronger connections for your devices [6].

In essence, mmWave is poised to revolutionize the way we experience wireless communication. It assures extremely fast data rates and allows you to connect to the internet, watch high-definition movies and participate in online gaming sessions without buffering. It is as if one has a personal fast lane for data even during peak times. With mmWave, the wireless communication future is bright as more devices can be connected allowing us achieve an uninterrupted online experience [7].

Leveraging the existing spectrum below 6 GHz, sub-6 GHz 5G technology offers a foundational layer for 5G deployments. This frequency range prioritizes wider coverage and superior signal penetration through buildings, making it suitable for ubiquitous network access in urban and rural environments alike. In contrast, mmWave 5G utilizes frequencies in the much higher range beyond 24 GHz. While mmWave boasts exceptional data speeds, its range is significantly shorter and signal penetration through obstacles is limited. A more detailed comparison of these two 5G technologies is provided in the Table 3.2.

**Table 3.2** Comparison between 5G mmWave and sub 6 GHz technologies.

| Feature | mmWave | Sub 6 GHz |
|---|---|---|
| Frequency range | Above 24 GHz | Below 6 GHz |
| Coverage area | Short range | Wider coverage |
| Data transfer rates | Very high (1 Gbps+) | Moderate (100–700 Mbps) |
| Penetration & propagation | Poor | Better |
| Use cases | High-speed broadband | Mobile broadband, IoT |
| Infrastructure requirement | Dense network of small cells | Traditional cell towers |
| Device compatibility | Limited | Widespread |
| Cost | High | Lower |
| Latency | Very low | Low |
| Standardization | More complex | More mature |

### 3.5.3 Small Cells

A given geographical area called a cell with a macro BS serves number of user nodes. In case of massive number of users or subscribers, particularly in crowded locations with lots of devices, the macro BS with limited antennas and RF chains fails to serve all the nodes. This is where small cells come into play. They are like miniature versions of cell towers scattered throughout a city. Consider a busy mall with one access point for all the people. The Wi-Fi is difficult to use because everyone is trying to utilize it. With so many more access points (APs), 5G technology amounts to a total mall makeover. Incorporating multiple small APs make room for serving more people and thus enhance the network performance.

The network capacity is increased by these tiny cells working in coordination with the bigger towers, or macrocells, already in place. The fact that 5G makes use of specific high-frequency radio waves makes them more crucial. These waves are like powerful spotlights instead of floodlights, carrying a lot more information across shorter distances. By having these small cells spread around, the 5G signal stays strong and reaches more devices, even in busy city centres [1, 8, 9].

Small cells provide a number of advantages. They provide blazingly high data speeds first. They can manage a lot more devices and data traffic at once, which eases congestion and keeps everything running smoothly for everyone. Thirdly, because they consume less electricity than conventional towers, they are more energy-efficient. This results in a more environment friendly network and prolongs the life of the phone's battery.

The reach of small cells is shorter than that of macrocells. For this reason, just like a team passing the baton in a relay race, they must cooperate as a network.

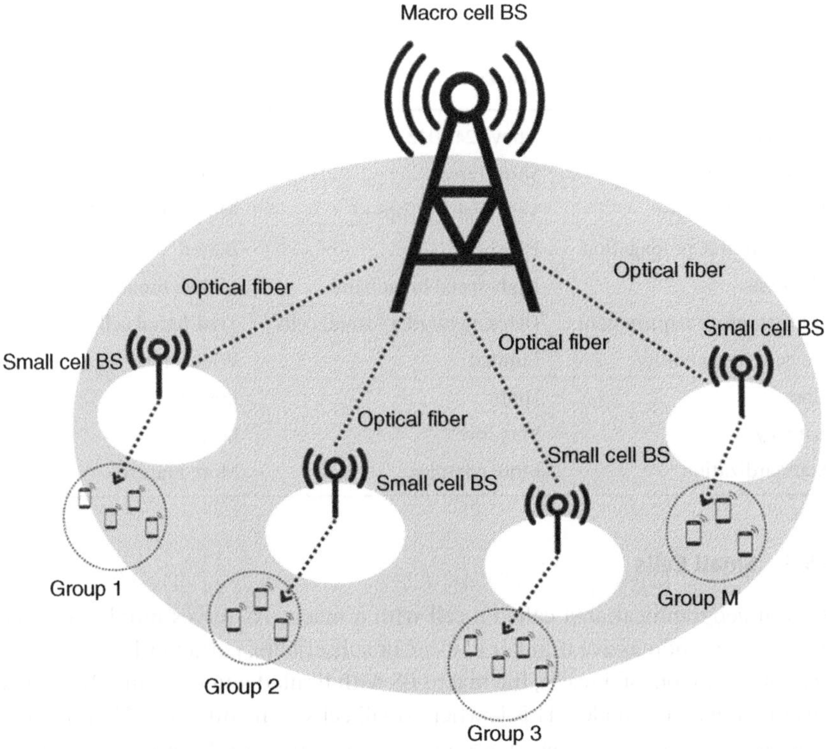


**Figure 3.6** A macro cell partitioned into various small cells.

No matter where you are in the city, you will always have a solid connection thanks to this so-called ultra-dense network [10, 11]. Figure 3.6 depicts the system model incorporating small cells.

### 3.5.4 Visible Light Communication

Visible light communication (VLC) is a way to transmit information using light or optical signals. It is a promising optical wireless communication (OWC) technology in which light signals are used for data transmission and reception. Apart from illuminating the room, the ordinary light bulb also transmits data to the devices at the same time as shown in Figure 3.7. That is the potential of VLC. This futuristic technology uses the very same light from light sources to transmit information, like a hidden code carried on a light wave. Instead of radio waves, VLC relies on light-emitting diodes (LEDs), the same efficient light sources found in many bulbs today. These LEDs can flicker on and off at incredibly fast rates, undetectable to the human eye, but carrying encoded messages. A special detector on your device

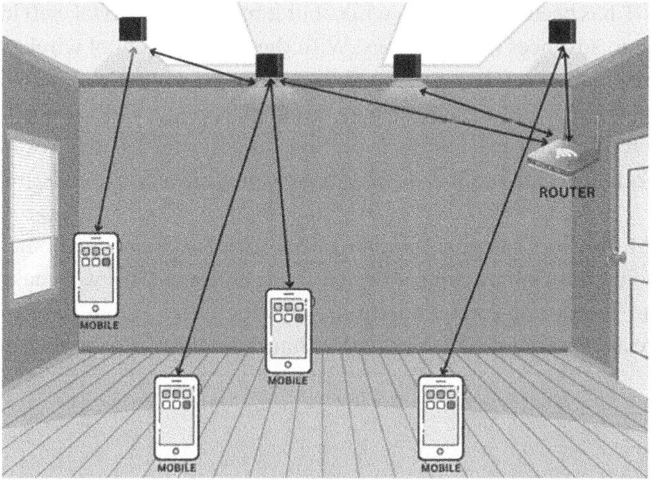


**Figure 3.7** An indoor communication scenario supported by VLC.

picks up these rapid changes in light, translating the hidden message back into usable data. The receivers are equipped with photodiodes (PDs) that receive the light signals which are decoded into useful data.

VLC offers a solution in situations where traditional radio waves fall short. Hospitals, for example, rely heavily on wireless communication for critical data transmission. But radio waves can interfere with sensitive medical equipment. VLC, on the other hand, would not disrupt these vital machines, providing a safe and reliable alternative for data transfer within hospitals. Additionally, airplanes rely on radio waves for communication, but at high altitudes, these signals can weaken or become unreliable. VLC has the potential to provide a more stable and secure communication channel for airplanes, enhancing safety and efficiency in the sky. VLC is still under development, but it holds the promise of super-fast internet in places where traditional radio waves are jammed or unreliable. It is like having your own personal light-based internet connection, all without needing extra cables or bulky equipment [12].

## 3.6 IoT and its Evolution

IoT is simply a large network of connected devices which communicate independently. IoT involves smart sensors, intelligent devices, connected terminals that collect the data and send it to one central hub. After that, this data can be utilized for a wide range of purposes, from improving our quality of life to resolving significant issues in fields like traffic control and healthcare.

The concept of IoT has been around for a while, but it really started to take off in recent years thanks to advances in technology. With the advancement of wireless technology, the large amount of data generated from IoT devices can be handled efficiently. The emergence of 5G, the next generation of wireless technology, is one of the main forces behind IoT and its evolution. In order for IoT devices to interact with each other, 5G offers substantially faster communication speeds and decreased latency.

The future of IoT involves massive number of IoT devices, from our homes and cars to our bodies and clothes, are connected to the internet. This will make it possible for us to gather and evaluate data in ways that were before impractical, which will result in new discoveries and enhancements to our quality of life [12, 13].

## 3.7 Role of 5G in IoT

5G serves as a superhighway for the vast amount of data that traverses the IoT. Millions of gadgets talking to one another all the time is how IoT is best defined. 5G is built to handle enormous amount of traffic that frequently overpowers traditional networks. Allowing smooth information exchange and connectivity between all these devices is the main function of 5G. 5G provides high data rates, enhanced network coverage and improved capacity with efficient resource utilisation and optimum power consumption thus yielding an improved IoT ecosystem. 5G combines these to provide a large and fast network that is perfect for the many needs of various IoT devices [12]. The conceptualization of a 5G IoT architecture with layers can be described in the way outlined. Figure 3.8 shows the five layered architecture of 5G IoT.

- **Physical Device Layer (IoT Sensors):** This layer is made up of the real-world sensors and gadgets that gather data. These 5G IoT gadgets would be made to talk to each other using 5G protocols.
- **Network Layer (IoT Gateway):** The IoT Gateway connects the devices to the 5G network. In case it's needed, it can translate protocols and send the data to the core network from different sensors.
- **Communication Layer (5G Link):** This layer is the 5G cellular network, which connects the gateway to the cloud and sends data quickly and with little delay.
- **Architecture Layer (Intelligence and Cloud Storage):** This layer includes the cloud platform and any processing power at the network edge that goes with it. This is where you can store data, analyse it, and maybe even make decisions in real time.

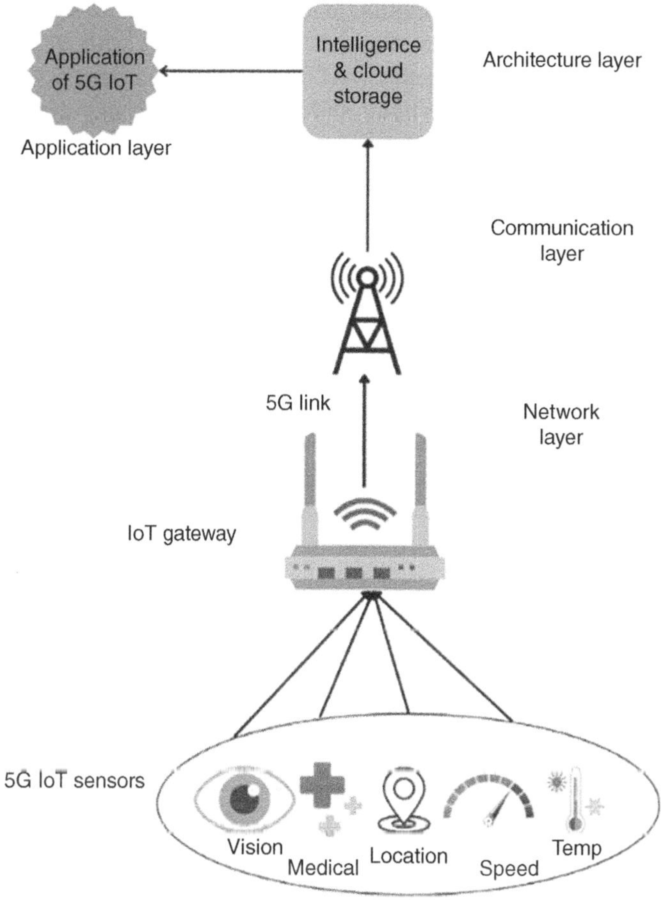


**Figure 3.8** Five layered architecture of 5G IoT.

- **Application Layer (Applications of 5G):** This layer is made up of all the different services and apps that use the data that is received and processed in the 5G IoT network. These could be smart city apps, industrial automation systems, remote health tracking, or any other service that benefits from high bandwidth and real-time data.

## 3.8 Integration of 5G IoT with Other Technologies

The real power of 5G IoT becomes evident when it smoothly combines with other cutting-edge technologies. 5G IoT is like a strong central nervous system that is perpetually gathering and transmitting data from numerous connected devices. However, the volume of data would be overwhelming without the analytical prowess of artificial intelligence (AI) and machine learning (ML). AI/ML algorithms form this ecosystem's brain, shifting through data for patterns, anticipating problems and suggesting best courses of action. It could be as simple as a smart thermostat, which learns your daily program and sets the temperatures appropriately or may be more complex; such as an industrial sensor that detects a slight tremor in a machine and predicts possible failure before it happens.

The extensive memory of this system is substituted by cloud computing. Presumably, 5G has a high data transfer rate such that huge quantities can be uploaded to the cloud for future processing and analysis. Compared to earlier systems, cloud computing replaces massive memory banks. As a result of fast data transfer speeds of 5G, data can be sent in large amounts to the cloud for more processing. Therefore, these powerful servers are used in big data analytics where they uncover patterns which cannot occur at device level. In fact, thousands of traffic sensors across town could stream real-time information into the cloud. Consequently, AI-based intelligent use of these materials can realize improved traffic flow while reducing congestion since it would have optimized traffic light settings. Current state-of-the-art research in 5G and its integration with other technologies is summarized in Table 3.3.

**Table 3.3** Current state-of-the-art research in 5G and its integration with other technologies.

| Ref. | Convergence with other technologies | Key performance indicator (KPI) | Summary |
|---|---|---|---|
| [10] | mmWave 5G | System capacity | This focuses on the challenges and opportunities presented by mmWave communication in small cells for 5G networks. It is observed that mmWave offers a significant advantage in terms of increased system capacity and address the spectrum limitations. |
| [17] | IoT, cloud computing | Latency, data rate | This paper explores a concept called CloudThings architecture, which aims to bridge IoT and cloud computing. It highlights the benefits of both IoT and Cloud computing. IoT allows objects to connect and share data with the web, while Cloud computing offers on-demand and scalable computing resources. |

*(Continued)*

**Table 3.3** (Continued)

| Ref. | Convergence with other technologies | Key performance indicator (KPI) | Summary |
|---|---|---|---|
| [18] | Edge computing, fog computing | Energy efficiency | The paper explores the potential of AI in the edge computing and fog computing regime. It highlights reinforcement learning techniques as particularly promising for making sequential decisions in resource management. |
| [21] | Digital twin, IoT | Power consumption, battery life | A novel approach to powering a massive IoT network with limited battery life of consumer devices is proposed in this paper. The key challenge addressed is the frequent battery replacements or charging required by these devices. |
| [23] | Digital twin, AI and ML | Latency | This paper explores a new technology called Network Digital Twin (NDT) designed for 5G and future networks. NDT creates a virtual representation of a physical telecom network, enabling control and synchronization without directly interacting with physical components. AI and ML are seen as crucial for this technology as it aims for reduced network latency. |
| [24] | LoRaWAN, 5G | Security, system efficiency | In this paper, the integration of LoRaWAN with 5G technology is presented. An extensible authentication protocol (EAP) is proposed that aims to enable the integration and provides secure access. The system performance and efficiency improve with the proposed protocol. |
| [25] | AI, 5G | Latency, data rate | This paper has highlighted the integration of AI and 5G for smart healthcare. The capabilities of AI involving efficient data analytics and decision making is coupled with robust 5G communication support. The proposed architecture reduces the computational cost by 20% and improves the system performance by 30%. |
| [26] | Digital twin, B5G | Latency, computational time | This paper proposes a placement strategy for battery constrained IoT devices in the edge network using digital twin. The latency is found to be 1.4 times lower than in the traditional approach. |

### 3.8.1 AI/ML

AI and ML stand as pillars in reshaping modern cities, where their integration revolutionizes various sectors. Think about a city filled with vitality where there are numerous traffic lights that have to adjust themselves according to the ever-changing traffic patterns. This is where 5G and AI/ML converge to create a smarter world. Acting as an information superhighway, 5G carries vast amounts of real-time data from traffic cameras, road embedded sensors and even connected vehicles. But without AI/ML, it would be like a disorganized bunch of numbers. The urban transport control centres are being replaced by AI/ML algorithms. They analyze real-time data streams for patterns and trends. For example, sudden unexpected rise in the number of vehicles on a specific route or an incident blocking off the busy intersection can be identified by AI. With this information at hand, AI can make decisions that optimize traffic flow. It can change timings for traffic lights on the fly, reroute cars away from congested roads, or send timely alerts to motorists when they anticipate delays ahead. This is like having a very smart conductor who understands what's happening everywhere in his district immediately he gets there.

The advantages go beyond alleviating the traffic jam only. By analyzing the past traffic flow, AI may be able to predict future traffic patterns. Such knowledge is useful to urban designers who would otherwise not know where to improve in their infrastructure such as constructing more lanes or reorganizing public means of transport. Furthermore, AI has the capability of using traffic data to determine areas that are prone to accidents and can introduce measures for safety purposes. Imagine a system that can respond, anticipate and prevent other future problems apart from current transportation issues [14].

This powerful combination of 5G and AI/ML is not limited to just traffic management. It can be applied to various scenarios. In a smart factory, AI can analyze sensor data from machines to predict potential maintenance issues, preventing costly downtime. For example, in a smart farm, weather data and soil conditions can be analyzed by AI so as to reduce irrigation and fertilizer use making way for greater crop yields. Therein lies boundless prospects where 5G provides fast data connection while AI/ML works as an intelligent engine at analyzing this information and ensuring sustainable development which ultimately becomes a smarter world economically [15].

### 3.8.2 Cloud Computing

Cloud computing provides on-demand access to computing resources and storage over the internet, eliminating the need for local servers or personal hardware. Suppose you have this big box with everything you hold dear to your heart, papers, photos, films and music. This box is just like your own personal storage space but it does not occupy any physical room instead it exists in the "cloud." The cloud is nothing more than renting a huge space in this gigantic virtual warehouse. As long as there is an internet facility around you, you can access all your stuff from anywhere anytime. Additionally, just imagine that this cloud can be used for storing complex mathematical computations and programs which are used for running various applications that we use daily. In the context of 5G, cloud computing offers a promising solution for supporting big data and its storage. In operation, 5G is like an ultrafast information highway. Big volumes of data moving at amazing rates can be handled by it. Dealing with the cloud makes this essential since it makes it possible to access and use all data and programs kept there practically instantly. Imagine an automobile that can drive itself. Several complex technologies are at work here, including sensors and microcomputers that are usually mounted directly on-board such vehicles, to navigate the road, identify objects, or make quick decisions to stop or veer out of a crash course. But, with 5G and cloud computing, the automobile can access these applications and do the computations in the cloud. The car will become lighter and more effective as a result of this liberation of space and computing power. Moreover, the car can access the most recent maps, traffic statistics, and safety procedures because the cloud can continuously update these applications with the most recent information. It is like having a super-smart team of engineers working behind the scenes all the time to enhance performance in the car.

Cloud computing with 5G is not only for self-driving cars. Think about a surgeon performing surgery remotely. The doctor could be miles away but with 5G high speed and low latency, they can get real-time data from medical devices. This allows for more precise and efficient surgeries, potentially saving lives [16]. The possibilities are limitless. Cloud-based systems can allow farmers to analyze soil conditions and optimize crop yields. To further customize their products and provide better customer service, companies can also save enormous volumes of consumer data in the cloud. At last, gamers have something to benefit from cloud gaming, which allows them to play intense games without the need for costly gaming consoles [17, 18].

### 3.8.3 Fog Computing

Fog computing is the approach that speeds up and lowers latency by moving data processing closer to the source of data creation. Imagine a city full of autonomous cars. Future homes will depend for safety on dynamically changing data. Fog computing and 5G combine here to guide this technological dance as cars navigate through traffic, respond to road conditions, and make snap decisions in a matter of seconds. Along with cameras and Lidar systems, there are sensors built within the cars that are always absorbing data about the surroundings. All of these are connected with the least amount of delay and at the highest speed feasible over a

5G network. However, sending all that data to a central cloud computer would take too long. It can be a matter of milliseconds deciding between an accident waiting to happen or a perfect swerve. This is when fog computing comes in handy, standing like an angelic guard at the edge of the network. In essence, fog computing places mini-cloud computers strategically throughout cities. Fog nodes stand closer to where information originates from- autonomous cars themselves. With 5G super-fast connection, data from the cars is instantaneously transmitted to the nearest fog node. The real power lies in the fog node's ability to analyze data in real-time. The fog node is capable of carrying out real-time data analysis because it has strong processing capabilities. Consequently, the fog node comprehends traffic flow and ascertains possible dangers, and then computes the best course for each car. It might mean adjusting speed, changing lanes or even initiating emergency braking procedures, all taking place in milliseconds.

The fog node should be thought of as a local control centre for individual self-driving cars. It has access to current information about immediate areas and can make snap decisions that are life-saving. This removes the necessity to depend on a distant cloud computer which means instantaneous decision-making which is critical for safe and efficient autonomous driving. Not only self-driven vehicles benefit from this development. Fog computing aided by 5G may change various industries completely. In a smart hospital, patient's real-time data from wearable sensors can be processed through fog nodes to give faster diagnosis or customized treatment plans. Imagine surgeons wearing augmented reality glasses with relevant patient information sent by nearby fog nodes, leading to more precise and effective surgical procedures. Fog Computing and 5G together make an edge computing powerhouse. Closer than ever before they bring processing power to the source of data enabling instantaneous decision-making through real-time data [19, 20].

### 3.8.4 Digital Twin

The concept of digital twins is fast becoming popular in the constantly changing world of technology. A powerful technology that makes a virtual replica of a physical item, always learning and developing itself with its original copy is digital twin. That's exactly what digital twins are all about, and when you combine them with the high-speed capabilities associated with 5G, they open up endless possibilities.

At its core, a digital twin is an intricate computer model. This model perfectly copies every single inch of the physical object. Nevertheless, unlike an inert blueprint, the 5G-powered digital twin comes to life through data handling. Real-time data from sensors inserted into anything from jet engines to concert halls to entire buildings is continuously transmitted at blazing rates over the 5G network to their online copies. Using strong algorithms and simulations to examine incoming sensor data, it then forecasts possible issues and maximises performance. Consider an airplane engine that has witnessed a technical fault or vibration, for instance. This anomaly is provided by the digital twin, in which a virtual replica of airplane engine is created and diagnosed for fault tolerance. It can also be used to forecast its progress and hence support early maintenance before a failure happens. This kind of vision has revolutionized industries with sophisticated machinery as it prevents costly downtimes and guarantees smooth operations. Going digital with twins has numerous benefits, Predictive Maintenance being simply one. After then, twins can also maximize performance.

The digital twin, through analysis of historical data and simulation of different scenarios can come up with ways to enhance physical object's efficiencies. For example, the wind farm's digital twin might suggest changes in turbine settings to maximize energy production. This would mean big savings as well as reduced environmental impact.

The prospects for both 5G technology and digital twins are immense. Imagine a city having its own digital twin that constantly analyzes real-time information from such things like weather stations, traffic cameras and power grids. Consequently, the system may decide on how best traffic flow should occur when breakdowns are expected or even include some plans concerning whether to have new energy-efficient infrastructures in place. It is more or less having an ever-learning and changing monitor for your physical world that will make everything run smoothly and efficiently in future where data dominates [21, 22].

#### 3.8.4.1 Digital Twin Lifecycle: From Data to Transformation

The digital twin, a virtual representation of a physical entity, goes through a well-defined lifecycle with distinct stages that leverage data to optimize the physical counterpart as shown in Figure 3.9. Here's a breakdown of these stages:

1) Connect and Collect:
   This is the foundation stage. Sensors and other data collection methods are attached to the physical entity, gathering real-time or historical data on its performance, operation, and environment. This data could include temperature readings, vibration levels, energy consumption, or user behaviour, depending on the specific application.
2) Learn and Analyze:
   The collected data is loaded into powerful analytics tools and ML algorithms. These algorithms examine the data to look for trends, patterns, and potential issues. The digital twin "learns" and gains a deeper understanding of the behaviours and capabilities of the real entity by using this data.

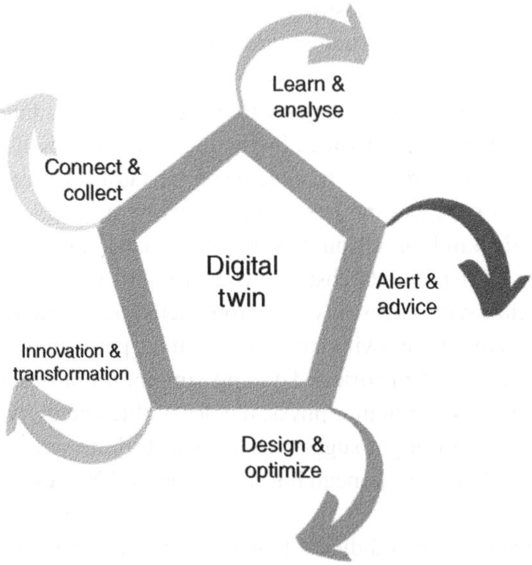


**Figure 3.9** Digital twin.

3) Alert and Advise:
   The digital twin can produce notifications for any issues or inefficiencies based on the analysis. For instance, it might pick up on a deviation in sensor readings that points to a potential equipment malfunction. Additionally, it can provide proactive guidance on performance optimization, like recommendations for changing operating parameters.
4) Design and Optimize:
   The digital twin develops into an effective optimization and simulation tool. Engineers can build virtual models and test various scenarios using the data and insights from the digital twin. This reduces risks and maximizes efficacy by enabling them to assess possible changes (such as design alterations or maintenance techniques) prior to putting them into practice on the physical entity.
5) Innovation and Transformation:
   The digital twin stimulates innovation in addition to enhancing current systems. New goods, services, and business models can be created using the information and understanding gathered throughout the course of the lifetime. Based on real-world usage data, a digital twin of a manufacturing facility, for instance, could spot chances for process automation or the development of whole new goods. It is not always a linear lifespan. The constant processes of data collecting and analysis feed back into the stages of learning and optimization. The ultimate objective is to produce a dynamic, ever-evolving digital representation that enables the physical object and the systems it interacts with [23].

## 3.9 Techniques to Improve the Performance of Wireless Networks

A lot of techniques have to be observed to optimize performance and ensure a smooth user experience given the advent of 5G, which brings with it ultra-fast data speeds and massive network capacity. Here are four key techniques.

1) User Selection/User Scheduling:
   Consider a scenario of a crowded restaurant having few number of waiters. They can't possibly serve everyone at once, so they have to prioritize. This is how user selection in 5G networks operate. The network has to select some devices over others for data transmission when there are many devices wishing for connection.

   5G employs intelligent algorithms that take into account factors such as signal strength, data quality and UE capabilities. For example, if someone is downloading a big video file they could be prioritized over someone else who is looking at their email for an uninterrupted viewing experience; this dynamic selection ensures effective utilization of system resources while each user's requirement is met.
2) Channel Estimation:
   In a noisy room, communication encounters challenges akin to those addressed by channel estimation in 5G networks, where obstacles and interference can distort radio wave signals as they traverse through the air. These distortions necessitate accurate channel estimation to be provided to the network for effective communication. The base station and UE use methods such as pilot signals and channel reciprocity to estimate radio channel characteristics in 5G. With this, the network can change its power levels, transmission parameters like coding schemes amongst others to ensure reliable data transfer. Basically, this means that while you were trying your best to communicate clearly while speaking in that clamorous room, its volume and clarity were being adjusted by the network [27].

Figure 3.10 shows a block diagram of a wireless system. The channel estimator module is used to estimate the channel characteristics using the received signal. Using the estimated channel state information, the transmitted data/information is extracted from the received signal. There are various channel estimation techniques used in the wireless communication system. Channel estimation algorithms play a crucial role in 5G networks by reducing the distortions and interferences that radio signals encounter while being transmitted. Figure 3.11 shows the classification of different channel estimation techniques. These techniques act as detectives by examining the received signal to ascertain the

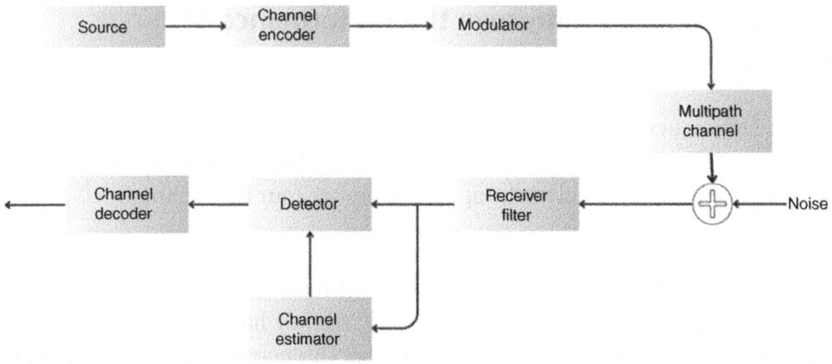


**Figure 3.10** Wireless communication model depicting channel estimation.

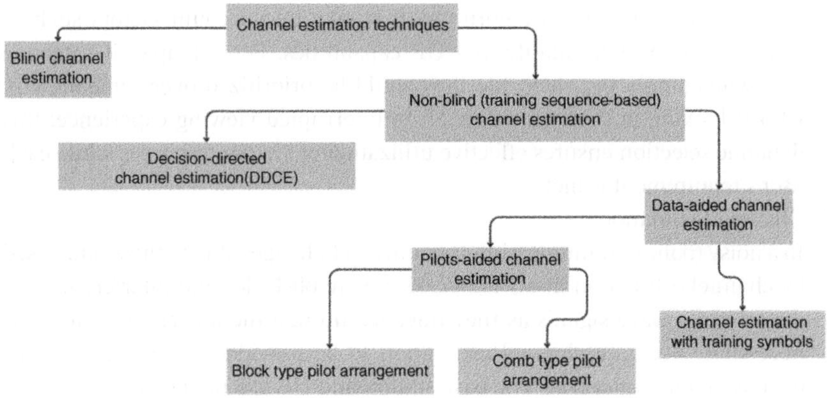


**Figure 3.11** Channel estimation techniques.

characteristics of the channel without any prior knowledge about the channel. A synopsis of various significant methods is provided below:

1) Blind Channel Estimation:
   As its name implies, this technique works without the usage of pilots or other pre-embedded training messages. All that is considered in estimating the channel are the statistical characteristics of the signal received and the provided data.
2) Non-Blind Channel Estimation (NCE):
   This category estimates the channel using a restricted amount of information about the transmitted signal. Many NPN-NCE techniques are in common use:
   - Decision Directed Channel Estimation (DDCE):
     DDCE, or Decision-Directed Channel Estimation, uses the correlation between the received signal and its derivative to estimate the channel.

- Data-Aided Channel Estimation:

This large group tries to guess the channel by using data that is sent with it. Here are some common ways to do it:

- **Pilot-Aided Channel Estimation**: A lot of people like this method because it sends data along with known pilot signals. The network can get a general idea of the channel by seeing how these tests change when it is present. A number of methods are available for hiring pilots, including:
  - **Block-Type Pilot Arrangement**: Block-Type Pilot Arrangement means that during a broadcast frame, pilots are sent to all subcarriers at special times. For channels that fade slowly, which means that their qualities change slowly over time, this method works well.
  - **Comb-Type Pilot Arrangement**: To set up a frame in the shape of a comb, pilots are put into specific subcarriers of each sign. For channels that fade quickly, which means that their properties change quickly, this way works better.
- **Channel Estimation with Training Symbols**: Using known training symbols instead of pilots to figure out the channel is called channel estimation with training symbols. For example, pilot-aided estimation is used, but it may use different training routines.

By employing these various channel estimation techniques, 5G networks can effectively decipher the "noise" introduced by the channel and recover the original transmitted data. This ensures clear and reliable communication, even in challenging environments [27].

3) Energy Optimization:

It is often seen that the smartphone's battery drain out quickly while streaming a film. The latter has been addressed in relation to energy optimization in 5G networks. There is a greater level of energy efficiency for 5G networks as opposed to earlier ones. Methods like physically partitioned and low giving devices sleep phases have been used. Network slicing enables network operators to realize virtual networks within their physical infrastructure that can be tailored specifically for different applications with different energy requirements whereas device sleep modes allow UE to enter low power modes when not actively transmitting or receiving data, thus significantly extending battery life. As such, these characteristics provide an equilibrium between network performance and user device battery lifetime.

4) Resource Management:

In a 5G network, resource management is like a well-organized tool box in which all tools have their own place where they fit. Resource management in 5G operates on this same principle. The resources used by 5G include the radio spectrum, power and bandwidth among others. Proper distribution of these resources ensures that user needs are met efficiently.

Intelligent algorithms in 5G networks allocate resources selectively depending on prevailing traffic and user demands. Thus, it allows for capacity optimization of the system so as to facilitate several users and applications simultaneously on-board. This is similar to when network intelligently distributes out tools of different tasks hence ensuring everyone has what one requires to perform the network tasks effectively.

## 3.10 Performance Parameters of Next Generation Wireless Systems

This section unveils the performance metrics of next generation wireless systems, capacity, spectral efficiency, and energy efficiency. As we delve into the realm of next-generation wireless systems, like 5G and beyond, three key performance parameters take center stage: capacity, spectral efficiency, and energy efficiency. Each plays a critical role in shaping the user experience and determining the overall effectiveness of these advanced networks. Let's embark on a detailed exploration of these parameters, uncovering their technical nuances and practical implications.

1) **Capacity**: The power to serve
Consider a busy traffic road during the peak hours of usage of cars in the city. For wireless networks, capacity refers to their ability to meet increasing needs of users and equipment. In essence, it shows how much data can be transmitted over an area in a given period. This implies the number of users that a network can accommodate at once and still maintain an acceptable level of service quality. Technically, capacity is measured using bits per second (bps) or Megabits per second (Mbps) for low data rates and Gigabits per second (Gbps) for high-speed applications such as ultra-high-definition video streaming. There are several factors affecting network capacity including:
**Radio Spectrum Bandwidth**: This term refers to how many frequencies are available for sending out data. The larger the bandwidth, the greater the potential capacity; a broader highway can support more lanes of traffic. 5G works with higher frequency bands compared to its predecessors, unlocking a wider spectrum for increased capacity.
**Modulation Techniques**: These methods demonstrate the encoding of information on radio waves prior to transmission. 5G makes use of high-level modulation formats such as 1024-Quadrature Amplitude Modulation (1024-QAM), which increases the amount of data that can be carried by a signal by fitting several bits of information in one.

2) **Spectral Efficiency**: Extracting more from less
Another analogy that one can continue with is the highway analogy, which speaks to how efficiently we use the radio spectrum available. In essence, this measures how much data can be sent per bandwidth unit (bps/Hz). This metric becomes increasingly important as data demand continues to soar and spectrum resources become scarce. Normally, spectral efficiency is stated as bits per second per Hertz (bps/Hz). Here's how 5G tries to enhance spectral efficiency:
**Advanced Channel Coding**: The coding schemes for 5G utilize sophisticated codes like Low-Density Parity-Check (LDPC) codes that assist in efficient use of bandwidth allocated.
**Higher-Order Modulation**: As already mentioned above, higher-order modulation schemes like 1024-QAM can send more data bits per symbol thereby maximizing the information carried within the given spectrum.
**Resource Allocation Techniques**: These techniques allocate resources such as bandwidth and power dynamically according to the current traffic situation. Thus, it ensures that the spectrum is used efficiently by avoiding situations where some areas have less traffic than others leading to unused capacity.
3) **Energy Efficiency**: Powering the future
Consider a marathon runner attempting to save energy for a protracted event. The capacity of wireless networks to provide expected services with the least amount of energy possible is the definition of their power efficiency. This becomes a crucial statistic when the number of devices and data traffic rises. Good networks, for instance, equate to cheap operator operating costs and extended battery life for consumer devices. It is technically measured in bits per Joule (bps/J) that 5G aims at increasing its energy efficiency:
**Network Slicing**: In this method, operators can use network slicing technique which enables them to select several virtual networks within one physical network with every slice built for separate applications based on different energy needs. For instance, there can be a network slice dedicated to IoT low-power devices which can work more efficiently than those slices serving high-speed data download requirements.
**Device Sleep Modes**: UE when not transmitting or receiving data actively, could be placed in low-power states such as sleep mode which significantly reduces power consumption at the UE side accordingly.
**Beamforming Techniques**: These techniques focus radio signals towards specific UE instead of broadcasting them omnidirectionally. This reduces energy wasted on unnecessary transmissions and optimizes power usage [28].

### 3.10.1 The Elaborate Rhythm of Performance Indicators

The ability to carry traffic, the rate at which information is carried per unit bandwidth, and amount of useable energy are not separate things. There is usually a complicated relationship between them. For example, more energy consumption may be used when we increase capacity through wideband or high order modulation. Likewise, the amount of coding intended to maximize spectral efficiency would determine how much data could be transmitted. The trick is to strike an appropriate balance between these benchmarks based on specific network requirements or deployment scenarios [29].

These performance indicators will evolve as we move forward toward next generation wireless systems.

## 3.11 Challenges and Future Directions

Despite its promises, 5G has challenges to address. This section will probe into the major constraints that must be resolved for full unlocking of 5G's potential and explore future directions that this technology is bound to take.

*Challenges to Address*:

**Investment in Infrastructure**: Developing a strong 5G network requires a significant infrastructure expenditure. Specifically, especially in crowded places, new cell towers must be equipped with sophisticated antenna systems (massive MIMO). Moreover, the shorter range of the high-frequency bands used by 5G makes a network of tiny cells necessary to guarantee uniform coverage. Telecom companies must therefore make substantial initial outlays in building out vast infrastructures [30].

**Device Compatibility**: Currently, not all devices can handle what 5G offers. To fully enjoy speed and low latency benefits of 5G, consumers would need to upgrade their smartphones and other devices. Until then there may be temporary inequality in access [30].

**Security Matters**: The implementation of a novel technology such as 5G generates additional security concerns. The increasing volume of data traffic and the necessity for a more extensive network infrastructure may present an opportunity for cyber-attacks. Consequently, it is imperative that we establish robust security protocols that will enable us to maintain the reliability and safety of our 5G networks [31].

**Energy Cost**: The increased power consumption associated with 5G's advanced features, such as minuscule cells, underscores the importance of developing energy-efficient solutions. In order for 5G advancements to be sustainable, it is imperative that end-user devices and network infrastructure be optimized for energy efficiency [31].

*Future Steps*:

**Network Segmentation**: This new innovation offers an opportunity for network slicing wherein it is possible to create virtual networks within a physical one. This facilitates customization according to specific needs. Just think about it- one slice could be dedicated to self-driving cars requiring ultra-reliable low-latency connectivity while another slice would be needed by virtual reality applications demanding high bandwidths. Network slicing unlocks a future of tailored network experiences.

**Integration with AI**: The combination of 5G's capabilities with AI promises significant advancements. Real-time data collected from a multitude of connected devices can be analyzed by AI to optimize network performance, predict maintenance needs, and even personalize user experiences.

**The Rapid Rise of IoT**: 5G is a perfect catalyst for the explosion of IoT. It paves way for smart cities, connected industries and remote healthcare advancements through connecting a huge number of devices with low latency. Imagine wearables persistently watchful over vital signs or censors in buildings that automatically adjust temperature and lighting, 5G fosters a future of interconnected intelligence.

**Unexpected Innovations**: Like any disruptive technology, 5G can lead to unexpected innovations. The combination of ultrafast speeds, zero delays, and extensive network capacity may open doors to entirely new applications and services that are beyond our imagination.

By overcoming these challenges and embracing such exciting future directions, the potential for 5G to revolutionize how we connect, interact and experience things around us is limitless. This chapter has equipped you with knowledge about the insides of 5G, its present challenges as well as its transformative possibilities for what lies ahead.

## 3.12 Conclusion

The role of 5G technology in addressing the needs of massive future IoT networks is highlighted in this chapter. The capabilities of 5G include high data rate, low latency, high spectral efficiency and high energy efficiency which serves multitude of IoT nodes. The 5G technologies of MIMO, massive MIMO, small cells and mmWave communication provide extended network coverage and high capacity. The VLC communication offers illumination and data transmission capabilities within the existing lighting infrastructure. The 5G can effectively converge with other technologies like AI, ML, cloud computing, fog computing and digital twin for a more robust IoT ecosystem. The performance of 5G IoT can be enhanced with techniques which includes resource allocation, energy optimization, user selection and channel estimation. In conclusion, 5G networks represent a paradigm shift in wireless communication, offering ultra-high data rates, low-latency communication, and seamless IoT integration for hyper-connectivity era ahead. Despite challenges, concerted efforts in research and collaboration can overcome hurdles and unlock the transformative potential of 5G. As this journey progresses, striving for a future where connectivity drives progress, innovation, and societal well-being is imperative.

## References


1. Shehab, M.J., Kassem, I., Kutty, A.A. et al. (2021). 5G networks towards smart and sustainable cities: a review of recent developments, applications and future perspectives. *IEEE Access* 10: 2987–3006.
2. Ahmadzadeh, S., Parr, G., and Zhao, W. (2021). A review on communication aspects of demand response management for future 5G IoT-based smart grids. *IEEE Access* 9: 77555–77571.
3. Sanchez-Gomez, J., Carrillo, D.G., Sanchez-Iborra, R. et al. (2020). Integrating LPWAN technologies in the 5G ecosystem: a survey on security challenges and solutions. *IEEE Access* 8: 216437–216460.
4. Madani Fadoul, M., Nour Hindia, M.H.D., Dimyati, K. et al. (2020). Oblique projection: an interference mitigation for MIMO multihop full-duplex relay. *Transactions on Emerging Telecommunications Technologies* 31 (4): e3799.
5. Larsson, E.G., Edfors, O., Tufvesson, F., and Marzetta, T.L. (2014). Massive MIMO for next generation wireless systems. *IEEE Communications Magazine* 52 (2): 186–195.
6. Saluja, D., Singh, R., Saluja, N., and Kumar, S. (2022). Connectivity improvement of hybrid millimeter wave and microwave vehicular networks. *IEEE Transactions on Intelligent Transportation Systems* 24 (2): 1456–1464.
7. Qamar, F., Hindia, M.N., Abbas, T. et al. (2019). Investigation of QoS performance evaluation over 5G network for indoor environment at millimeter wave bands. *International Journal of Electronics and Telecommunications* 65 (1): 95–101.
8. Chopra, G. (2023). A review on rate splitting multiple access: challenges and future directions. In *2023 International Conference on Emerging Smart Computing and Informatics (ESCI)*, Pune, India (01-03 March 2023), pp. 1–7. IEEE.
9. Shafique, K., Khawaja, B.A., Sabir, F. et al. (2020). Internet of things (IoT) for next-generation smart systems: a review of current challenges, future trends and prospects for emerging 5G-IoT scenarios. *IEEE Access* 8: 23022–23040.

**10** Liu, C., Li, M., Hanly, S.V. et al. (2018). Millimeter-wave small cells: base station discovery, beam alignment, and system design challenges. *IEEE Wireless Communications* 25 (4): 40–46.

**11** Shen, B., Lei, Z., Huang, X., and Chen, Q. (2018). An interference contribution rate based small cells on/off switching algorithm for 5G dense heterogeneous networks. *IEEE Access* 6: 29757–29769.

**12** Shahinzadeh, H., Mirhedayati, A.S., Shaneh, M. et al. (2020). Role of joint 5G-IoT framework for smart grid interoperability enhancement. In 2020 15th International Conference on Protection and Automation of Power Systems (IPAPS), Shiraz, Iran (30-31 December 2020), pp. 12–18. IEEE.

**13** Dhaka, A., Nandal, A., and Dixit, R. (2021). Cognitive radio network-based design and security challenges in 5G communication. In: *Research Anthology on Developing and Optimizing 5G Networks and the Impact on Society*, 135–150. IGI Global.

**14** Rizvi, S. (2023). Unveiling the potential of artificial intelligence and machine learning in the 5G network landscape: a comprehensive review. *Asian Journal of Research in Computer Science* 16 (4): 23–31.

**15** Hamdi, W., Ksouri, C., Bulut, H., and Mosbah, M. (2024). Network slicing based learning techniques for IoV in 5G and beyond networks. *IEEE Communications Surveys & Tutorials* 26 (3): 1989–2047.

**16** Khoda, M.E., Razzaque, M.A., Almogren, A. et al. (2016). Efficient computation offloading decision in mobile cloud computing over 5G network. *Mobile Networks and Applications* 21: 777–792.

**17** Zhou, J., Leppanen, T., Harjula, E. et al. (2013). Cloudthings: z common architecture for integrating the internet of things with cloud computing. In Proceedings of the 2013 IEEE 17th international conference on computer supported cooperative work in design (CSCWD), Whistler, BC, Canada (27-29 June 2013), pp. 651-657. IEEE.

**18** Iftikhar, S., Gill, S.S., Song, C. et al. (2023). AI-based fog and edge computing: a systematic review, taxonomy and future directions. *Internet of Things* 21: 100674.

**19** Basir, R., Qaisar, S., Ali, M. et al. (2019). Fog computing enabling industrial internet of things: state-of-the-art and research challenges. *Sensors* 19 (21): 4807.

**20** Meng, Y., Naeem, M.A., Almagrabi, A.O. et al. (2020). Advancing the state of the fog computing to enable 5g network technologies. *Sensors* 20 (6): 1754.

**21** Taneja, A. and Rani, S. (2024). Energy efficient digital twin enabled massive iot network with use case in consumer health. *IEEE Transactions on Consumer Electronics* 70 (1): 2099–2106.

**22** Ramirez, R., Huang, C.Y., and Liang, S.H. (2022). 5G digital twin: a study of enabling technologies. *Applied Sciences* 12 (15): 7794.

**23** Mozo, A., Karamchandani, A., Sanz, M. et al. (2022). B5GEMINI: digital twin network for 5G and beyond. In NOMS 2022–2022 IEEE/IFIP Network Operations and Management Symposium, Budapest, Hungary (25-29 April 2022), pp. 1–6. IEEE.

**24** Jradi, H., Nouvel, F., Samhat, A.E. et al. (2023). A seamless integration solution for LoRaWAN into 5G system. *IEEE Internet of Things Journal* 10 (18): 16238–16252. https://doi.org/10.1109/JIOT.2023.3267502.

**25** Pradhan, B., Das, S., Roy, D.S. et al. (2023). An AI-assisted smart healthcare system using 5G communication. *IEEE Access* 11: 108339–108355. https://doi.org/10.1109/ACCESS.2023.3317174.

**26** Chukhno O., Chukhno N., Araniti G. et al., Placement of social digital twins at the edge for beyond 5G IoT networks. in *IEEE Internet of Things Journal*, vol. 9, no. 23, pp. 23927–23940, 2022, https://doi.org/10.1109/JIOT.2022.3190737.

**27** Pourkabirian, A. and Anisi, M.H. (2021). Robust channel estimation in multiuser downlink 5G systems under channel uncertainties. *IEEE Transactions on Mobile Computing* 21 (12): 4569–4582.

**28** Xue, Q., Liu, Y.J., Sun, Y. et al. (2022). Beam management in ultra-dense mmWave network via federated reinforcement learning: an intelligent and secure approach. *IEEE Transactions on Cognitive Communications and Networking* 9 (1): 185–197.

**29** Brilhante, D.D.S., Manjarres, J.C., Moreira, R. et al. (2023). A literature survey on AI-aided beamforming and beam management for 5G and 6G systems. *Sensors* 23 (9): 4359.

**30** Abdelkader, G., Elgazzar, K., and Khamis, A. (2021). Connected vehicles: technology review, state of the art, challenges and opportunities. *Sensors* 21 (22): 7712.

**31** Zikria, Y.B., Kim, S.W., Afzal, M.K. et al. (2018). 5G mobile services and scenarios: challenges and solutions. *Sustainability* 10 (10): 3626.

**32** Zhang, L., Zhao, H., Hou, S. et al. (2019). A survey on 5G millimeter wave communications for UAV-assisted wireless networks. *IEEE Access* 7: 117460–117504.

# 4

# Emerging Communication Paradigms for 6G IoT: Challenges and Opportunities

*Aditya Soni[1], Ashu Taneja[1], Neeti Taneja[2], and Laith Abualigah[3]*

[1]*Chitkara University Institute of Engineering and Technology, Chitkara University, Punjab, India*
[2]*Department of Computer Science Engineering, Noida Institute of Engineering and Technology, Greater Noida, Uttar Pradesh, India*
[3]*MEU Research Unit, Middle East University, Amman, Jordan*

## 4.1 Introduction

From the birth of first generation (1G) analog systems to the current deployment of fifth generation (5G), mobile communication networks have evolved through generations, ushering in a disruptive era with previously unheard-of capabilities. In order to fulfill the demands of a society that is becoming more and more data-driven, research attention is shifting toward the potential of sixth generation (6G) communication systems as 5G networks mature globally. Pioneering developments are anticipated in these systems. The landscapes of convergent communication, localization, and sensing systems are explored in detail in this thorough investigation, revealing the complex interactions between technological enablers, difficulties, and potential future developments.

The voyage begins with an emphasis on how 6G promises to transform wireless communication paradigms, highlighting the significance of important elements like localization and sensing techniques. While addressing important research topics, the article analyzes potential enabling technologies, evaluates environment-aware applications, and suggests trends. Recognizing the potential of 5G new radio (NR) access interfaces that provide significant bandwidth, high carrier frequencies, and large antenna arrays, the study predicts that 6G systems will follow a path that leads to even higher frequencies, even wider bandwidths, and cutting-edge technologies like terahertz (THz) and millimeter waves (mmWave) communication.

As the exploration deepens, the narrative seamlessly transitions to the imperatives of supporting massive IoT applications in the 6G era. A comprehensive survey

*Current and Future Cellular Systems: Technologies, Applications, and Challenges*, First Edition.
Edited by Garima Chopra, Suhaib Ahmed, and Shalli Rani.

unfolds, outlining drivers, requirements, network architectures, breakthrough technologies, and use cases for 6G-enabled massive IoT. The limitations of existing 5G networks become evident, setting the stage for the impending 6G era to address challenges spanning data-rate, latency, reliability, and global coverage.

Continuing the journey, emerging applications such as collaborative robots, holographic telepresence, deep-sea tourism, autonomous driving and Internet of Everything (IoE) emerge as catalysts for the development of 6G. A detailed survey outlines societal and technological trends, requirements, key enabling technologies, ongoing research projects, and standards efforts, providing a holistic view of the current developments towards 6G.

Figure 4.1 shows the technological advancement from 1G to the current deployment of 5G that marks a transformative journey, with the upcoming 6G era promising groundbreaking advancements. 6G will aim to revolutionize the IoT applications to offer fully autonomous and intelligent future systems. The future 6G-IoT applications are also depicted in Figure 4.1 which include internet of healthcare things, satellite internet of things, internet of medical things (IoMT), internet of drone things (IoDT) and vehicular internet of things. 6G aims to offers massive connectivity, high data rates, reduced latency, ultra-high reliability and efficient resource optimization with reduced computational overhead. The comparison of the wireless technological evolution is shown in Table 4.1 on the basis of various system parameters.

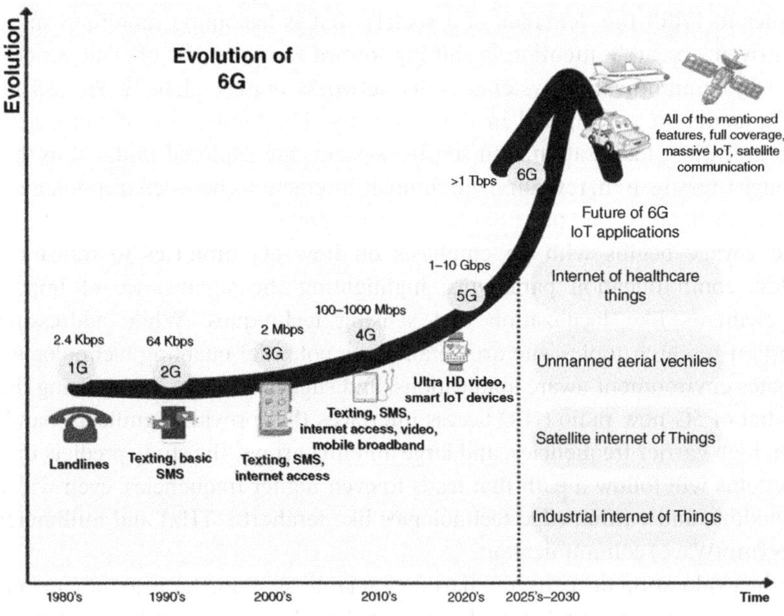


**Figure 4.1** Technological evolution from 1G to 6G and role of 6G in various IoT applications.

**Table 4.1** Comparison of wireless technologies on the basis of different system parameters.

| Parameter | 2G | 3G | 4G | VoLTE /VoWiFi | 5G | 6G |
|---|---|---|---|---|---|---|
| Technology | GSM/1991 | WCDMA/2001 | LTE, WiMAX/2009 | IMS/2012 | MIMO, Massive MIMO, Co-Operative MIMO, HetNets, mm Waves | Advanced MIMO, AI Integration, Terahertz Communication |
| Advantages | Digitally encrypted in between mobile and cellular base station | Diminished overloading in actual systems | Global access, scalable mobile service | Improved coverage and Superior call quality | Enhanced transmission speed and lower latency | Extreme high-speed data transfer, Ultra-low latency |
| Frequency | 1.8 GHz | 1.6–2 GHz | 2–8 GHz | HD Voice up to 7 kHz, Full-HD Voice 20 Hz to 20 kHz | 3–30 GHz | 100 GHz and higher |
| Data Rate | Up to 64 Kbps | 2 Mbps speed for static and for mobile users up to 384 Kbps | supports data rate up to 1 Gbps for immobile users and up to 100 Mbps for mobile users | 50 Mbps | support data rate > 1 Gpbs | Multi-Terabits |
| Bandwidth | 1 Mbps | 2 Mbps | 20 Mbps to 1 Gbps | 1 Gbps | 1 Gbps & higher | Multi-Terabit |
| Challenges | Encryption, Mobile Computing | Security and privacy, QoS | Security, Integration of IP Devices | Circuit-switched fallback, IMS profile VoLTE | Cost, Coverage and Deployment | Sustainability, Environmental Impact, Interoperability with legacy systems |

The exploration takes a comprehensive turn towards the future, delving into the state of the art and vision of 6G. This chapter paints a vivid picture of the 6G landscape including emerging technologies, use case scenarios, key performance parameters, drivers, novel architectures and technology enablers. The integration of artificial intelligence, blockchain, tactile Internet, and emerging technologies becomes a focal point, guiding the reader through the potential evolution and challenges.

Conclusively, the amalgamation of these diverse introductions provides a panoramic overview of the journey from 5G to the imminent 6G era. The narrative weaves through technology enablers, application scenarios, research trends, and the broader societal and industrial landscape, offering a cohesive understanding of the multifaceted path towards the future of wireless communication.

### 4.1.1 Breakthrough 6G Technologies

#### 4.1.1.1 Holographic MIMO (Multiple Input Multiple Output)

Holographic MIMO is an emerging wireless technology that uses the principle of holography to improve the efficiency of the wireless systems. Using an array of guided antennas, the beamforming is enabled which is a signal processing approach that minimizes the angular range while transmitting radio signals accurately in targeted directions. Improved network efficiency, reduced interference, higher signal-to-interference noise ratio (SINR), improved coverage, high throughput, user tracking and localization are just a few advantages of this strategy.

A new technique called holographic beamforming (HBF) uses holograms to guide the beam through the antenna. By capturing the interference patterns between object wave and reflected wave, hologram is created and the signals are beamformed in a desired direction nullifying the interference from other directions. From the field distribution of reference wave and reflected object wave, hologram map is obtained according to the holographic principle shown in Figure 4.2. Using this technique, radio waves go to the antenna's back, scatter across its front and then reposit themselves to take on the intended beam direction and shape. HBF uses Software Defined Antennas (SDA), which are less expensive, heavier, smaller, and need less power than MIMO systems or conventional arrays [1].

#### 4.1.1.2 Intelligent Reflecting Surfaces (IRSs)

Intelligent reflecting surfaces are a promising wireless technology that offers versatile solution to the problem of limited signal coverage due to unfavorable propagation environment. Also, known as reconfigurable intelligent surfaces

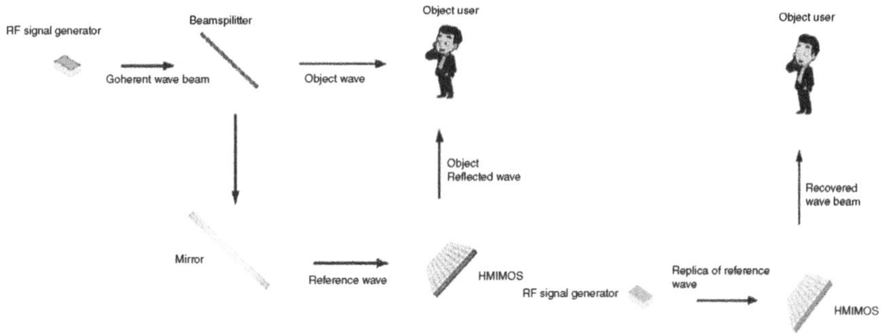


**Figure 4.2** Holographic principle.

(RISs), these enable seamless connectivity by intelligently controlling the propagation environment. These consist of large number of low cost passive reflecting elements whose reflection properties can be adjusted to enable smart radio. The incident signals can be beamformed in the direction of desired user by introducing sufficient phase shifts through IRS controller. Promising for 6G networks, IRSs provide proactive electromagnetic properties in the surrounding environment. IRS reduces the detrimental effects of antenna correlations and places less restrictions on antenna spreading than standard beamforming. Apart from passive IRSs, active IRSs have also gain popularity due to their amplification capability in addition to controllable reflection. When it comes to 6G, IRS can facilitate seamless network changes by allowing autonomous technology selection without requiring manual configuration. 6G networks promise to address handover problems, latency, and data losses reported in current cellular networks by shifting towards a cell-free approach and non-orthogonal communications through the use of intelligent and energy-efficient hybrid approaches, multiple access computing, and smart radio-based devices [1, 2].

#### 4.1.1.3 Cell free Massive MIMO

The concept of cell-free massive MIMO (multiple-input multiple-output) is mentioned in the discussion surrounding the transition in 6G networks from traditional cellular and orthogonal communications to cell-free and non-orthogonal communications. In cell-free approach, there are no cell boundaries and the user nodes in a network are served by multitude of access points (APs). With 6G, users will not need to manually configure networks instead, the network will select the most suitable communication technology based on their needs, enabling seamless network switching. The term "cell-free communication" describes this.

With this modification, issues with data loss, handover failures, and delays that arise during cell transfers in current networks are resolved. The edge nodes or the users in the shadowed areas are given service through distributed APs which cooperate among themselves to optimise the system performance. In contrast to one AP in the centre of the cell in massive MIMO system, the cell-free system allows multiple user nodes to communicate through a set of APs. This results in extended network coverage, high throughput, reduced latency through efficient resource allocation [1, 3].

#### 4.1.1.4 Edge Computing

Problems like high energy consumption and slow processing rates have emerged as a result of the fast growth of apps like augmented reality, virtual reality, and mobile gaming, which have put a strain on mobile devices' computational power. In order to solve these problems, cloud computing and edge computing have emerged as potential solutions, with the aim of moving processing from mobile devices to more powerful servers. Much attention has been focused on edge computing's ability to relocate processing power closer to the data source, which reduces latency and improves system performance overall. This paradigm shift is crucial in the era of digital society and smart cities, when real-time data processing is becoming more and more necessary [4].

Edge computing involves a network of edge nodes, edge devices, cloud servers, and network infrastructure. By processing data locally, edge computing reduces network latency and bandwidth use while enhancing data security and privacy. However, there are also disadvantages, such as heterogeneity in devices, resource limitations, and reliability and availability of the system.

A task-oriented triple-stage monitoring-based high-reliability edge-side mobile terminal shared computing architecture is proposed as a remedy for these issues. This architecture integrates mobile device processing capability, monitors task completion, identifies faults, and ensures dependable and efficient task completion [5].

#### 4.1.1.5 Terahertz (THz) Communication

There is a growing interest in wireless technology running at very high frequencies, such as THz in anticipation of 6G networks. THz has even more bandwidth potential than millimeter wave (mmWave) technology, which makes it appealing for use in next communication systems. THz offers benefits like tolerance to atmospheric effects and the capacity to enable ultra-high performance in throughput and reliability, especially in contexts like indoor communication and virtual reality (VR)/augmented reality (AR) applications, despite drawbacks such significant path loss. Massive connectivity is promised by THz technology, which also makes highly integrated systems with hundreds of antennas possible. But putting THz gear into practice is still difficult, particularly when it comes to effectively modulating signals. In spite of these obstacles, THz technology appears to have a bright future in next-generation wireless networks as research and development continue.

#### 4.1.1.6 Quantum Communication

Along with molecular communication, one of the emerging concepts in communication is quantum communication. The sentence emphasizes how crucial it is to plan for 6G networks in order to satisfy the constantly expanding needs for wireless communication. It implies that a great deal of research is being done on a number of 6G network-related topics, such as cutting-edge projects and enabling technologies. Among the technologies that are expected to form the foundation of 6G are visible light communication (VLC), THz communication, and quantum communication. The goal of the paper is to spur academics to investigate these 6G system enabling technologies and their IoT applications [6].

## 4.2 Internet-of-Things and its Evolution

The network of physical items or things integrated with sensors, software, and other technologies to connect and exchange data with other devices and systems over the internet is known as the Internet of Things, or IoT for short. These objects can be anything from commonplace things like wearable technology and home appliances to large-scale industrial machines and infrastructure elements.

The primary goal of IoT is to empower these items to gather and share data on their own, resulting in more automation, convenience, and efficiency across a range of domains and sectors of the economy. Users can get real-time updates and insights, remotely monitor and operate these devices, and even automate actions based on preset circumstances or algorithms by connecting these devices to the internet.

IoT has the potential to revolutionize a wide range of industries, including manufacturing, retail, energy management, healthcare, agriculture, education, and environmental monitoring. IoT facilitates better decision-making, resource efficiency, and general quality of life by establishing smart environments where gadgets converse with humans and each other. By streamlining procedures, increasing output, and providing individualized service, it transforms sectors and supports environmental preservation and sustainability initiatives. IoT creates a future where intelligence and connectedness combine to radically change how we work, live, and interact with the world around us. It does this by being able to gather and analyze massive volumes of data in real-time, which opens up countless potential for innovation and advancement.

### 4.2.1 Role of 6G IoT

With the goal of expanding human activity to extreme environments such as high altitudes, space, oceans, and deep sea areas, a vision has emerged for a comprehensive and interconnected system that includes earth, sea, sky, and space. Driven by the Internet of Everything (IoE) concept and enhanced by edge computing, this system is proposed by 6G as a four-tier architecture consisting of space, air, ground and underwater/sea networks.

The space class includes various satellites, especially very low Earth orbit (VLEO) satellites, providing high data rates, low latency, and accurate localization, especially to serve remote areas and difficult locations. At the aerial level, unmanned aerial vehicles (UAVs), balloons, and balloons are represented as aerial mobile systems, providing flexibility by approaching ground IoT devices to collect data and serve as computing hubs.

The ground level refers to traditional wireless networks such as cellular and local networks, including advances such as terahertz communication between 6G targets. Because water has different propagation characteristics from land, the underwater/sea level uses optical communications to provide Internet services to dispersed nodes.

6G places a strong emphasis on edge computing as crucial computing, providing decentralized services and real-time operations to satisfy the demands of IoT-enabled applications. This method uses edge intelligence and machine learning to enhance the administration of Internet of Things systems.

Furthermore, 6G promises more than just improved perception; it also highlights distribution as a critical component of large-scale IoT system futures. Concerns like data intensity, processing needs, latency sensitivity, privacy, and security must all be taken into consideration in the expanding IoT world at 6G. Real IoT data is produced by the switch from VR/AR to holographic and five-way communication. This growth in data necessitates higher latency requirements for apps, more processing and analysis capability, and increased focus on data security and privacy protection.

With 6G's wireless coverage and integration of technologies like smart computing, intelligent communication, user sensing and localization, data analytics, data processing, and automation, future IoT networks and applications stand to gain a great deal. The future's 6G networks, in contrast to 5G networks, should provide wide coverage and enhanced flexibility to support the networking environment and service delivery of the Internet of Things [7].

### 4.2.2 6G IoT Framework

The IoT ecosystem is anticipated to grow significantly as a result of the full-dimensional wireless coverage and functional integration that 6G networks will offer. In order to build an interconnected environment, the convergence of IoT and 6G entails linking a myriad of devices, sensors, actuators, smartphones, computers, and RFID systems. Without the need for human interaction, this integration attempts to facilitate seamless data sensing, processing, and

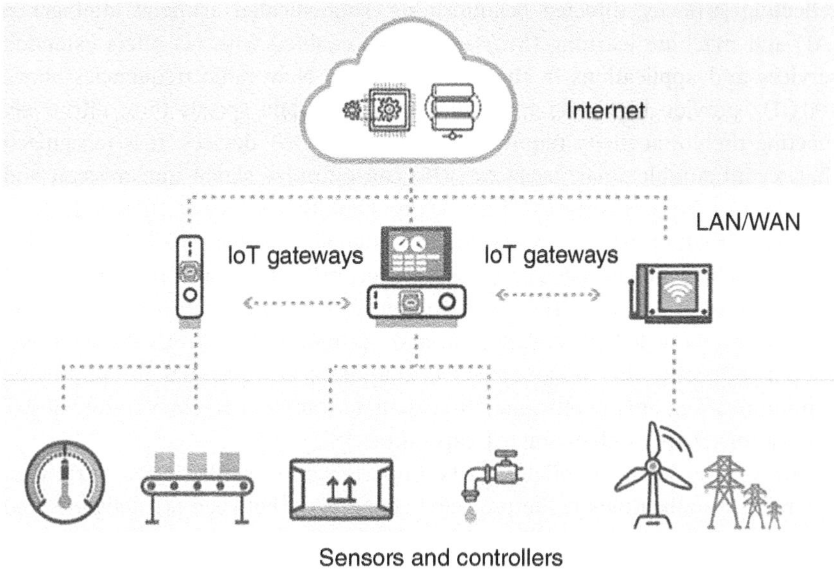


**Figure 4.3** A comprehensive 6G IoT framework.

communication. Massive coverage and improved adaptability are two features that the 6G networks are expected to deliver, supporting a wide range of IoT services and applications. By 2030, it is expected that there will be 500 billion Internet-connected objects, making 6G's ability to support IoT applications' needs even more crucial. A 6G IoT framework highlights how important it is to smoothly integrate IoT into the architecture of 6G networks. The convergence of IoT and 6G networks, as shown in Figure 4.3, is expected to transform connectivity by allowing a wide range of systems, sensors, and devices to be easily integrated. 6G's full-dimensional wireless coverage and greater adaptability are expected to play a significant role in allowing a wide range of applications across smart cities, healthcare, industrial automation, and linked autonomous vehicles. By 2030, 6G has the potential to support up to 500 billion IoT devices. It is anticipated that 6G with IoT will enable a variety of applications, from connected driverless vehicles and smart cities to healthcare and industrial automation [1].

## 4.3 Enabling 6G Technologies for IoT

This section presents the enabling 6G technologies that finds usage in the IoT sector. The capabilities of 6G communication systems can be integrated with the IoT evolution thus transforming the digital world. The introduction of intelligent reflecting surfaces, directed beamforming, sophisticated artificial intelligence (AI) and machine learning (ML) algorithms enabled with 6G offers extended services and applications in the IoT ecosystem. New radio frequencies above 100 GHz provide high data rates and significant data speeds thus, effectively meeting the connectivity requirements of various IoT devices. It is recognized that reconfigurable smart surfaces (RIS) can optimize signal transmission and reception in IoT networks by dynamically regulating the electromagnetic environment. Location-based services and the strategic placement of IoT devices are made possible by beamspace processing. It becomes clear that combining AI and ML approaches is essential for controlling device communication, processing large amounts of IoT-generated data, and optimizing IoT network operations. To put it briefly, the discussed 6G technologies are essential for improving communication, energy efficiency, and spatial awareness across a variety of IoT applications, hence enhancing IoT capabilities [8].

One major factor highlighting its importance in furthering a variety of IoT-related applications is the projected cooperation between 6G networks and IoE in the 2030s. An expansion of IoT, IoE integrates a number of components, including people, data, things, and processes. This integration will help advance IoT-related services by supporting large-scale machine-type communication and sensor devices. The proper management of a myriad of intelligent devices requires the smooth synergy between 6G and IoE, which demands features like fast data rates, low latency, and scalable connectivity. It emphasizes how the success of IoE depends on 6G's capabilities, highlighting how it helps connect a large number of intelligent devices and makes significant data collection possible for advanced analytics across a variety of areas. In general, it is projected that 6G convergence for IoE applications will transform the environment of many applications by offering unmatched connection and data processing power for a wide range of IoT-related technologies [9].

The conversation around 6G technology explores important factors that will shape the Internet of Things. By studying many technological factors, it highlights the crucial role that 6G is likely to play in improving IoT capabilities. Within this context, IoT is envisioned as a fundamental component of 6G networks, with the aim of connecting a diverse range of devices to the Internet. Through a variety of IoT devices such as sensors, smartphones, computers, and radio frequency identification (RFID) tags, a connected network is created in which data processing, sensing and communication take place seamlessly. Anticipated to be a key enabler for diverse future IoT verticles, the 6G network is poised to provide extensive wireless coverage, connected communication support and reduced latency. The progress in 6G is projected to offer substantial support for IoT connectivity and services, surpassing the capabilities of the current 5G mobile network. Furthermore, the integration of AI into 6G networks is underscored

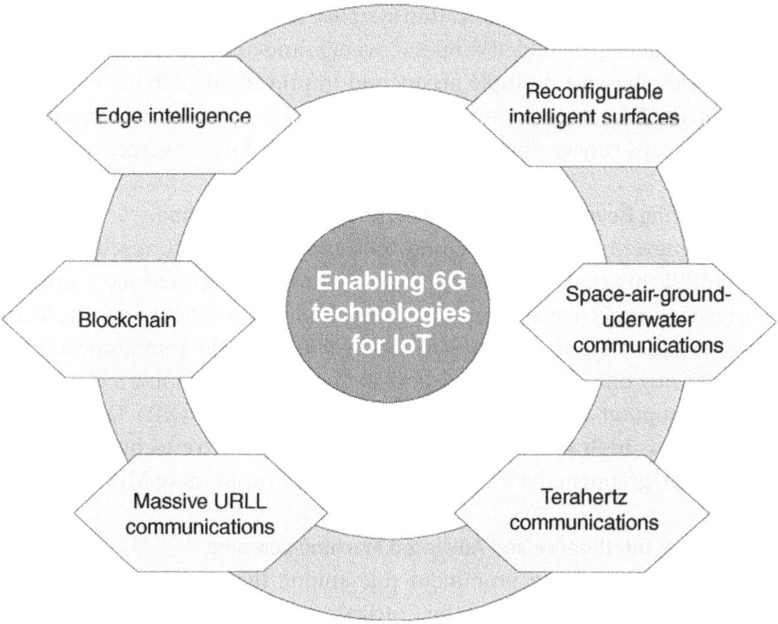


**Figure 4.4** Enabling 6G technologies for IoT.

as a crucial element in achieving self-sufficiency and intelligent systems. AI is expected to introduce full automation to 6G, enhancing the capabilities of radio signals and enabling cognitive radio transformations. The application of AI in real-time communications is considered pivotal for enhancing data transmission efficiency, network selection, handover procedures, and various other aspects. The technologies like digital twin, mobile edge computing, integrated sensing and communication, wireless information and energy transfer (WIET), dynamic network slicing, and other innovations contribute to the evolution of 6G and its impact on IoT capabilities. In summary, the comprehensive examination of these technologies signifies their integral roles in shaping the future landscape of IoT within the 6G era [1, 10, 11]. Figure 4.4 shows some of the enabling technologies of 6G for IoT for diverse communication regimes.

### 4.3.1 Convergence with Other Key Technologies

The previous section discusses the key 6G technologies that offer intelligent service capabilities in the IoT sector enabling automation, sensing, communication and computation. This section discusses the convergence with various other key technologies for future 6G communication systems. The segment on 6G enabling technologies explores several key advancements essential for the development of future 6G communication systems. As a significant IoT facilitator, 6G is expected to deliver extended wireless coverage and massive connectivity. The aim is to provide massive multiple access and improved adaptability compared to 5G, supporting IoT connectivity and services. Here is a condensed overview of the pertinent details concerning 6G IoT technologies and its convergence.

#### 4.3.1.1 Advancing Beyond Sub-6 GHz Towards THz Communication

6G aims to achieve data rates exceeding 1000 times that of 5G, reaching up to 1 Tbps. To fulfill this goal, exploration beyond sub-6 GHz frequencies is crucial. While early 6G systems may rely on sub-6 GHz and mmWave networks, the focus is shifting towards THz frequency band (0.1–10 THz). THz communication is vital for supporting ultra-high data rate applications like holographic telepresence (HT), Brain-Computer Interface (BCI), and Extended Reality (XR). Overcoming challenges such as high propagation losses require innovative technologies like nano-fabrication, graphene-based components, and ultra-Massive MIMO.

#### 4.3.1.2 Artificial Intelligence and Advanced Machine Learning

Artificial intelligence takes a prominent role among the enabling technologies for 6G, offering robust capabilities for intricate optimization tasks within the increasingly intricate and diverse mobile networks. Machine Learning (ML) is a data-driven tool that offers computational intelligence from the physical layer to network management. It can be divided into three categories: supervised, unsupervised, and reinforcement learning. Deep learning, a subset of ML, imitates biological nervous systems and finds applications in adaptive transmission, channel estimation, RF design, and intelligent network management. Advanced ML techniques, including federated learning and transfer learning, show promise in addressing data privacy concerns while enhancing wireless communications [12]. Beyond aiding network operations, AI extends its role to offer mobile AI services. This involves deploying deep edge resources to support computation-intensive AI tasks for various devices [13]. 6G envisions the integration of AI and ML at its core. The roles of AI and Federated Learning (FL) become pivotal in various wireless and mobile networking applications. FL, particularly, emerges as a distributed AI solution that prioritizes privacy by enabling devices to train local models and share updates rather than raw data. Data-driven service capabilities, distributed AI, support for connected communication, data security, privacy protection, and ease of large-scale deployment in 6G networks are some of the benefits of AI and FL.

#### 4.3.1.3 Compressive Sensing

A promising signal processing technique used to reconstruct the signals from smaller number of samples is the compressive sensing (CS). The data is efficiently acquired using CS without the need for high transmission bandwidth and data storage capabilities. It is a key enabler for 6G which uses signal sparsity to represent them with fewer coefficients thus reducing the computational overhead. It finds applications in image processing, radar communication, digital signal processing and medical imaging. Many communication systems use compressive sensing for massive machine type communication (mMTC) in 6G leveraging its potential in intelligent reflecting surfaces, non-orthogonal multiple access (NOMA) converged with deep learning.

#### 4.3.1.4 Blockchain/Distributed Ledger Technology

Blockchain technology plays an important role in 6G IoT ecosystem by offering capabilities of security and authentication [14]. Through the distributive ledger technology (DLT), it can prevent data breaches, unauthorized access thus maintaining data integrity. The stakeholders can manage the IoT data and its authentication through blockchain's cryptocurrencies techniques. It enables the IoT devices to transact values with each other autonomously. It offers dynamic resource allocation by enabling automation, for example, automated payments can be made for the IoT services. The decentralised nature of blockchain technology offers efficient decision making in contrast to the centralized model. The integration of blockchain powered by AI for the 6G IoT networks have been investigated guaranteeing data security [15]. To mange the device-to-device interactions, convergence with other technologies that includes edge computing, network slicing, network function virtualization (NFV), is also essential. Its applications span spectrum sharing, network virtualization, data storage and security in domains like UAVs, smart grids, smart healthcare and smart cities. The convergence of blockchain into 6G offers intelligent solution for network security, robust information exchange and flexible infrastructure support [14].

#### 4.3.1.5 Digital Twin

Digital twin technology, representing a logical copy of a physical object throughout its life cycle, emerges as a significant use case for 6G communication systems. In digital twin, the virtual replica of the physical entity is created and accessed to enable better decision making. Widely adopted in industries like manufacturing and aviation, digital twins utilize AI techniques to optimize processes. The application scope is anticipated to expand, particularly in healthcare and medicine, aligning with the full commercialization support. Challenges in scaling up digital twin platforms for millions of IoT devices and ensuring security and privacy must be addressed for broader deployment [16].

#### 4.3.1.6 Intelligent Edge Computing

The integration of edge intelligence (EI) in 6G plays a crucial role in enhancing network services and resource utilization. EI combines AI and ML techniques at the edge, automating tasks related to virtual resource management in NG-RAN architecture. Applications of EI extend to mission-critical services, massive mMTC, and energy-efficient solutions, offering opportunities for business models, varied industries to digitize the high-end applications in 6G mobile networks. Despite its advantages, there are number of challenges that require consistent efforts such as data inconsistency, insecure networks, resource scarcity, data privacy and model adaptability. 6G networks will integrate existing 5G and IoT infrastructures with edge computing hardware, enabling the execution of resource-intensive AI algorithms. An integral part of 6G computing is the mobility-enhanced edge computing (MEEC) that supports distributed large-scale clouds and facilitating effective computation at the edge.

#### 4.3.1.7 Dynamic Network Slicing

The process of dividing a network into several virtual networks, each suited to particular needs like performance, security, and dependability, is known as dynamic network slicing. Slices are virtual networks that are created, changed, and eliminated dynamically in response to the requirements of various services, applications, or users. It plays a very important role in 6G and offers consistent support to the users with the optimized delivery of services. The virtual networks into which a large network is sliced are optimally executed with improved performance. NFV and software defined networking (SDN) are the techniques required for implementation. It offers service customization, security, resource efficiency and scalability with guaranteed QoS.

#### 4.3.1.8 Big Data Analytics

Large data sets are examined using sophisticated ML and DL algorithms in order to optimise the decision-making capabilities. Big data is analyzed to find out the hidden information about market trends and customer needs. The management of big data sets and massive data is very crucial in 6G communication networks. The convergence of ML, DL, AI and big data is very useful in 6G networks to enable automation, reduced network overhead and latency [17]. The insights from the data are extracted using advanced data analytics for applications in varied domains like retail, manufacturing, finance, communication.

#### 4.3.1.9 Wireless Information and Power Transfer (WIPT)

The increasing energy consumption of the battery limited sensor devices and IoT nodes is the main challenge which demand for energy sustainability in IoT ecosystem. To overcome the challenge of increased energy overhead in the IoT landscape, the use of energy harvesting techniques is being promoted to increase the energy efficiency of the IoT network. WIPT is a promising energy conservation scheme that aims to harness energy from the radio frequency (RF) signals. Then, the harnessed energy is utilized for information transfer. Using energy from electromagnetic fields and waves, WIPT offers reduced energy overhead and thus is a ground-breaking 6G technology. With the adoption of WIPT in wireless networks, the frequent battery replacements can be minimised. Its potential lies in extending the battery charging lifetime of wireless systems, supporting devices without batteries, and enabling the progress of battery-less smart devices.

#### 4.3.1.10 Backscatter Communication

With backscatter communication, devices can interact with one another wirelessly by modulating or reflecting pre-existing signals rather than creating new ones. For use in 6G networks, ambient RF signals are employed in backscatter communication. In 6G networks, backscatter communication offers the possibility of increased network coverage, high throughput, and multiple access capabilities, resulting in better resource deployment and improved services. Together, these technologies set the course for 6G communication systems and offer solutions for a wide range of uses, including smart cities, healthcare, transportation, and other fields [18].

#### 4.3.1.11 Communication-Computing-Control Convergence

Communication, computation, and control are all expected to converge with 6G, turning the system into a complete computer. In this paradigm, wired connections in controlling systems are gradually replaced by wireless links, which are important. In order to maximize their performance, co-designing communication and control systems is stressed, highlighting the necessity of teamwork. To meet the demands of wirelessly connected industrial automation and the cloudification of mobile applications and networking services in 6G, communication, computing, and control must work together harmoniously. Advancements in mobile networks involve building on the strengths of previous architectures and incorporating additional benefits to meet evolving demands. As the next stage in this evolution, 6G networks are poised to capitalize on the achievements of 5G architectures while integrating innovative technologies to address future requirements. Various key technologies are pivotal in shaping the landscape of 6G communication systems, including the IoT, AI, WIET, MEC, UAVs, THz communications, dynamic network slicing, data analytics, holographic beamforming, cell free communication and backscatter communication.

In summary, 6G is anticipated to integrate a diverse range of technologies, encompassing THz communication, AI/ML, compressive sensing, blockchain/DLT, digital twin, edge computing, backscatter communication and WIET. These innovations are poised to address the requirements for ultra-high data rates, low-latency communication, and support emerging applications within the IoT ecosystem [9, 13]. The ever-evolving technological landscape prompts discussions about the integration of AI with mobile networks within the context of 6G. Envisioned as a vast computer, 6G is anticipated to integrate distributed resources encompassing communication, computing, storage, sensing, and control to accommodate pervasive computing, AI, blockchain, digital twin, and more.

## 4.4 Use Case Scenarios

### 4.4.1 Smart Healthcare

6G IoT transforms patient care in smart healthcare systems by enabling predictive analytics, individualized treatment plans, and remote monitoring. Healthcare professionals can respond proactively and give timely interventions by using wearable sensors and medical equipment to send vital signs and health data in real-time. This technology improves patient outcomes and saves lives by increasing accessibility, efficiency, and accuracy in healthcare delivery through telemedicine consultations and smart medication dispensers. As shown in Figure 4.5, the integration of 6G IoT in healthcare systems revolutionizes the way medical data is utilized and patient care is administered. The data from all the verticles of a healthcare unit are processed and it is ensured that all the verticles are in synchronization for efficient data exchange.

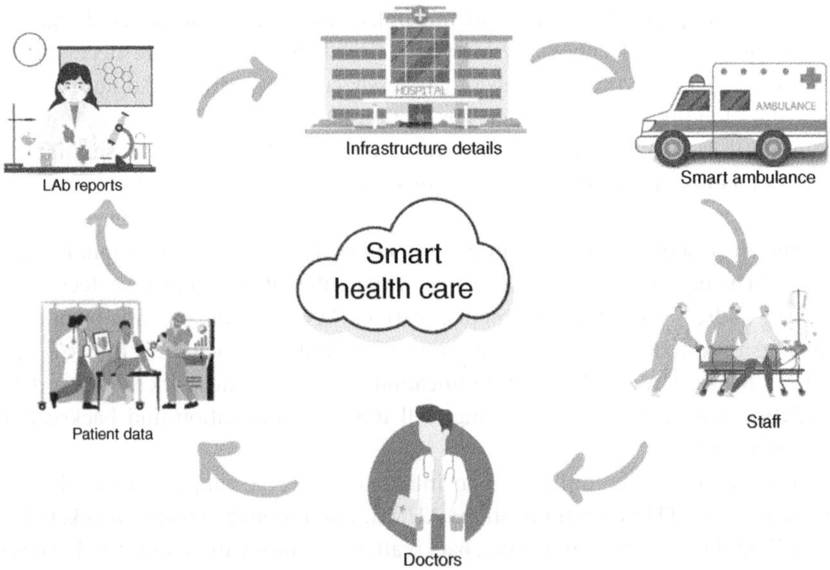


**Figure 4.5** Smart health care use case scenario.

### 4.4.2 Smart Transportation

With the advent of 6G IoT, transportation networks will be transformed into intelligent, seamlessly connected systems that improve accessibility, efficiency and safety. Vehicles may exchange real-time data about road conditions, traffic congestion, and possible hazards through ultra-fast 6G networks. The communication between vehicles is enabled through 6G powered vehicle-to-vehicle (V2V) communication. Also, the communication between vehicles and road side units (RSUs) or road side infrastructure, popularly known as vehicle-to-infrastructure (V2I) communication is also supported through 6G. This enables proactive navigation and collision avoidance. Furthermore, 6G-enabled sensors and intelligent ticketing systems installed in public transit networks improve routes and simplify passenger flow, lowering traffic and pollution while enhancing commuter satisfaction. Furthermore, the integration of shared mobility services and driverless cars made possible by 6G IoT promises to transform urban mobility and provide inhabitants with practical, environmentally friendly transit options [19]. A 6G powered smart transportation system is shown in Figure 4.6 which reflects connected vehicles support, fleet management, traffic management, route planning, travel assistance and vehicles' safety.

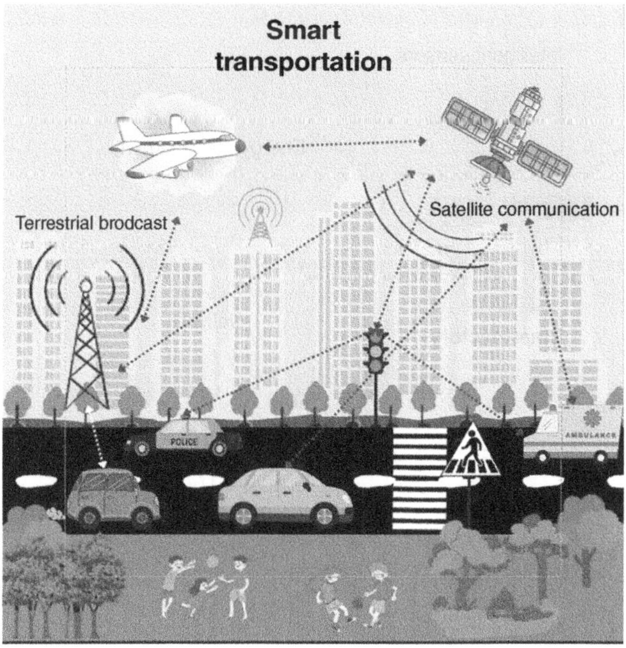


**Figure 4.6** An illustration of smart transportation system powered by 6G.

### 4.4.3 Smart Manufacturing

Throughout the production process, 6G IoT promotes sustainability, agility, and efficiency. Real-time monitoring of performance parameters is made possible by connected sensors integrated into machinery and equipment, which reduces downtime and allows for predictive maintenance. Sophisticated robots and self-governing systems enhance workflow automation, simplifying manufacturing and cutting expenses on labor. Manufacturers can maintain their competitiveness in the global market by streamlining supply chains, cutting waste, and reacting fast to shifting consumer needs with the use of machine learning algorithms and data-driven insights. Figure 4.7 shows a use case scenario of smart factory in which production, assembly, packaging and delivery is synchronized through intelligent and automated control using IoT. The inventory control, goods monitoring and supply-chain management is ensured through the use of intelligent sensors such that the operational efficiency is enhanced. The integration of advanced technologies facilitates a seamless and environmentally conscious approach to manufacturing, ensuring both economic viability and ecological sustainability.

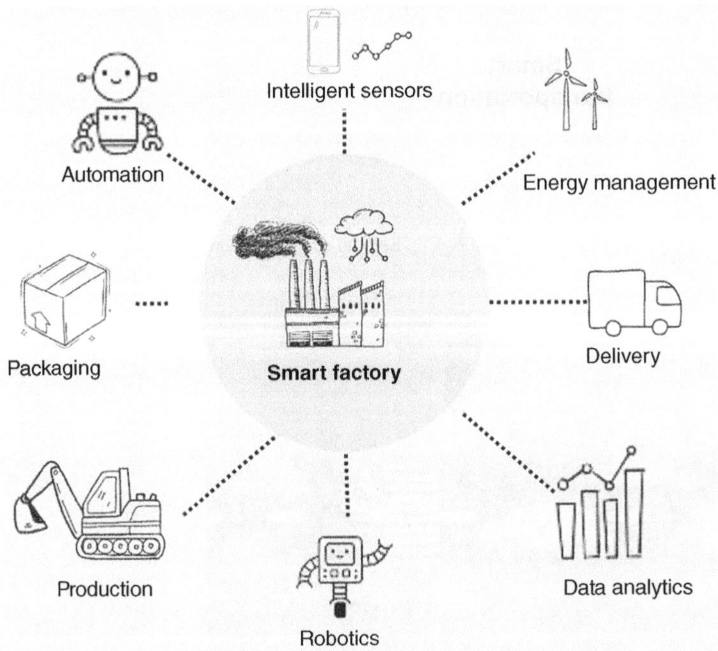


**Figure 4.7** A smart factory scenario managing production, assembly, packaging and delivery.

### 4.4.4 Smart Agriculture

Precision farming is made possible by 6G IoT, which enables a paradigm shift in the agricultural industry. Farmers can now make data-driven decisions that maximize crop output, reduce resource use and improve sustainability. Farmers can remotely operate irrigation, fertilization, and pest control systems with a network of sensors that continuously monitor soil moisture, nutrient levels, weather and crop health. Additionally, the efficiency and accuracy with which 6G-enabled drones and autonomous machinery can carry out activities like crop monitoring, sowing, and spraying will enhance productivity and decrease environmental impact. Figure 4.8 shows a smart agriculture scenario that makes use of IoT sensors to monitor the crop yield, soil fertility, water requirements, pest control and crop health in order to enable smart farming.

### 4.4.5 Smart Classrooms

Using 6G IoT, smart classrooms change the way that traditional educational models are taught, making them more participatory and individualized. Tablets,

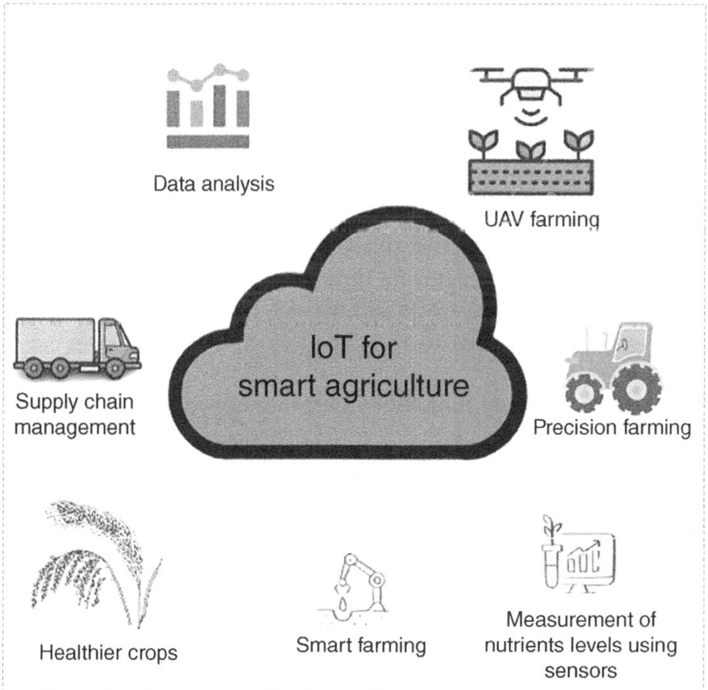


**Figure 4.8** Use of IoT in smart agriculture.

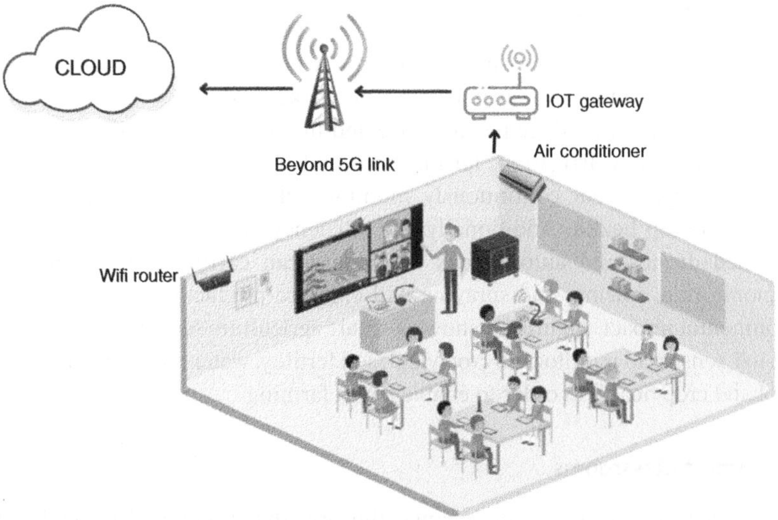


**Figure 4.9** Use case scenario of 6G IoT in smart classroom enabling smart learning.

interactive whiteboards, and educational apps are examples of connected gadgets that enable immersive learning environments where students can participate creatively with the course material. Teachers use data analytics to monitor their students' progress, customize their lessons to meet each student's needs, and create cooperative learning environments. Students acquire the critical thinking abilities necessary for success in the age of technology and obtain deeper insights via augmented reality simulations and virtual field trips. As shown in Figure 4.9, technological integration in education fosters interactive and personalized learning experiences, preparing students for the challenges of the digital era.

### 4.4.6 Smart Cities

6G IoT makes urban living revolutionary by enabling networked systems that improve sustainability, safety, and efficiency. Smart cities are able to optimize waste management, control energy consumption, improve traffic flow, and increase public safety by utilizing advanced sensor networks and real-time data analytics. 6G IoT transforms urban landscapes into responsive, adaptable settings that put residents' needs first while reducing their negative effects on the environment. Smart parking systems that lead cars to available spaces and intelligent street lighting that modifies brightness based on pedestrian activity are two examples of these environments [20]. The incorporation of such technologies turns cities into dynamic centers of advancement and invention, as illustrated in Figure 4.10.

**Figure 4.10** Smart city use case scenario.

## 4.5 Challenges Faced and the Solutions Offered

Even though 6G IoT has made promising strides, a number of issues still need to be resolved before its full potential can be reached. This section presents the challenges of 6G IoT and the suggested solutions:

1) An IoT network involves number of connected devices capable of exchanging data. The IoT nodes require reliable communication support to avoid delays or data loss. Thus, supporting large number of nodes in a network is challenge. 6G aims to support network scalability but the resource overhead and energy overhead with the increasing number of nodes is a bottleneck. Thus, the need is to design efficient network solutions and algorithms that allow the network to support large number of devices using minimum resources [21].
2) The millions of connected devices in a 6G enabled IoT network calls for strong measures for data security and privacy. The sensitive data is vulnerable to attacks, data breaches and unauthorized access [21]. Owing to massive devices, the huge amounts of data require strong encryption, authentication procedures, and privacy-preserving measures in place.
3) Since the IoT ecosystem involves number of diverse devices from different manufacturers, supporting different protocols and standards, it becomes very important to ensure/attain interoperatability. For integration of various IoT devices and enable seamless 6G communication, collaborative efforts should be encouraged among the network operators, manufacturers and other business models to offer interoperable solutions.

4) For diverse IoT applications including smart classrooms, smart healthcare, smart grids, industrial automation, intelligent transportation, 6G networks offer enhanced communication capabilities with low latency. To offer reliable communication support, it is necessary to converge the 6G with other communication technologies like blockchain, digital twin, cloud computing and AI for efficient faut tolerance and high network reliability.
5) Large data sets produced by IoT devices are examined using sophisticated ML and DL algorithms in order to optimise the decision-making capabilities. The management of big data sets and massive data is very crucial in 6G communication networks. The convergence of ML, DL, AI and big data is very useful in 6G networks to enable data management, data analytics offering enhanced operational efficiency.
6) The increasing energy consumption of the battery limited sensor devices and IoT nodes is the main challenge which demand for energy sustainability in IoT ecosystem. To overcome the challenge of increased energy overhead in the IoT landscape, the use of energy harvesting techniques is being promoted to increase the energy efficiency of the IoT network. Thus, the design of energy aware solutions enabling the integration of 6G and IoT is very important to achieve sustainable future networks [22, 23].
7) Another challenge faced by 6G IoT stakeholders is to comply with the regulations and ethical standards. It is important to adher to the rules and regulations pertaining to data ownership, privacy and accountability. For this, all the stakeholders should work together to offer compliance with the standard regulations.

## 4.6 Conclusion

This chapter presents the technological wireless evolution from 1G to 6G with emphasis on the 6G communication networks. 6G networks represent a paradigm shift in wireless communication, offering ultra-high data rates, low-latency communication, massive connectivity, improved network coverage and seamless IoT integration. The emerging 6G communication paradigms including IRSs, THz communication, backscatter communication, cell free massive MIMO are discussed. The role of 6G in the IoT landscape is investigated for connected communication support, reduced latency and enhanced operational efficiency. The convergence of enabling 6G technologies with AI, ML blockchain, digital twin, edge computing, offers intelligent solutions in energy optimised IoT networks. Further, the use case scenarios of 6G IoT in smart healthcare, smart transportation, smart agriculture, smart manufacturing, smart classrooms and smart cities are discussed. In the end, the benefits and challenges of 6G IoT are discussed. Despite challenges, concerted efforts in research and collaboration can overcome hurdles and unlock the transformative potential of 6G for a future where connectivity drives progress, innovation, and societal well-being for all.

## References


1 Qadir, Z., Le, K.N., Saeed, N., and Munawar, H.S. (2023). Towards 6G internet of things: recent advances, use cases, and open challenges. *ICT Express* 9 (3): 296–312.

2 Zeng, M., Bedeer, E., Li, X. et al. (2023). IRS-empowered wireless communications. In: *6G Wireless: The Communication Paradigm Beyond 2030*, 15. CRC Press, Taylor & Francis Group.

3 Basar, E. (2020). Reconfigurable intelligent surface-based index modulation: a new beyond MIMO paradigm for 6G. *IEEE Transactions on Communications* 68 (5): 3187–3196.

4 Pham, Q.V., Fang, F., Ha, V.N. et al. (2020). A survey of multi-access edge computing in 5G and beyond: fundamentals, technology integration, and state-of-the-art. *IEEE Access* 8: 116974–117017.

5 Ju, B., Liu, Y., Song, L. et al. (2023). A high-reliability edge-side Mobile terminal shared computing architecture based on task triple-stage full-cycle monitoring. *IEEE Internet of Things Journal* 10 (22): 20149–20161.

6 Manzalini, A. (2020). Quantum communications in future networks and services. *Quantum Reports* 2 (1): 221–232.

7 Giordani, M. and Zorzi, M. (2020). Non-terrestrial networks in the 6G era: challenges and opportunities. *IEEE Network* 35 (2): 244–251.

8 De Lima, C., Belot, D., Berkvens, R. et al. (2021). Convergent communication, sensing and localization in 6G systems: an overview of technologies, opportunities and challenges. *IEEE Access* 9: 26902–26925.

9 De Alwis, C., Kalla, A., Pham, Q.V. et al. (2021). Survey on 6G frontiers: trends, applications, requirements, technologies and future research. *IEEE Open Journal of the Communications Society* 2: 836–886.

10 Lu, Y. and Zheng, X. (2020). 6G: a survey on technologies, scenarios, challenges, and the related issues. *Journal of Industrial Information Integration* 19: 100158.

11 Giordani, M., Polese, M., Mezzavilla, M. et al. (2020). Toward 6G networks: use cases and technologies. *IEEE Communications Magazine* 58 (3): 55–61.

12 Vinuesa, R., Azizpour, H., Leite, I. et al. (2020). The role of artificial intelligence in achieving the sustainable development goals. *Nature Communications* 11 (1): 1–10.

**13** Janbi, N., Katib, I., Albeshri, A., and Mehmood, R. (2020). Distributed artificial intelligence-as-a-service (DAIaaS) for smarter IoE and 6G environments. *Sensors* 20 (20): 5796.

**14** Kumari, A., Gupta, R., and Tanwar, S. (2021). Amalgamation of blockchain and IoT for smart cities underlying 6G communication: a comprehensive review. *Computer Communications* 172: 102–118.

**15** Sekaran, R., Patan, R., Raveendran, A. et al. (2020). Survival study on blockchain based 6G-enabled mobile edge computation for IoT automation. *IEEE Access* 8: 143453–143463.

**16** Minerva, R., Lee, G.M., and Crespi, N. (2020). Digital twin in the IoT context: a survey on technical features, scenarios, and architectural models. *Proceedings of the IEEE* 108 (10): 1785–1824.

**17** Deepa, N., Pham, Q.V., Nguyen, D.C. et al. (2022). A survey on blockchain for big data: approaches, opportunities, and future directions. *Future Generation Computer Systems* 131: 209–226.

**18** Jiang, W., Han, B., Habibi, M.A., and Schotten, H.D. (2021). The road towards 6G: a comprehensive survey. *IEEE Open Journal of the Communications Society* 2: 334–366.

**19** Saluja, D., Singh, R., Saluja, N., and Kumar, S. (2022). Connectivity improvement of hybrid millimeter wave and microwave vehicular networks. *IEEE Transactions on Intelligent Transportation Systems* 24 (2): 1456–1464.

**20** Taneja, A., Saluja, N., and Rani, S. (2022). An energy efficient dynamic framework for resource control in massive IoT network for smart cities. *Wireless Networks* 1–12.

**21** Guo, F., Yu, F.R., Zhang, H. et al. (2021). Enabling massive IoT toward 6G: a comprehensive survey. *IEEE Internet of Things Journal* 8 (15): 11891–11915.

**22** Chopra, G. (March 2023). An efficient base station sleeping configuration for ultra-dense networks. *2023 International Conference on Emerging Smart Computing and Informatics (ESCI)*, pp. 1–5. IEEE.

**23** Rani, S. and Srivastava, G. (2024). Secure hierarchical fog computing-based architecture for industry 5.0 using an attribute-based encryption scheme. *Expert Systems with Applications* 235: 121180.

# 5

# Securing the Internet of Things: Cybersecurity Challenges, Strategies, and Future Directions in the Era of 5G and Edge Computing

*Geetanshi[1], Harshit Manocha[1], Himanshi Babbar[1], and Cherry Mangla[2]*

[1]*Chitkara University Institute of Engineering and Technology, Chitkara University, Punjab, India*
[2]*Assistant Professor, Concordia University of Edmonton, Edmonton, AB, Canada*

## 5.1 Introduction

The Internet of Things (IoT) comprises a networked computer system of real-world entities, encompassing programs, devices, and innovations. These entities seamlessly share and transmit data to other equipment via wireless connections [1]. The concept of the Web of Things has emerged as a pivotal technology in the contemporary era, enabled by the proliferation of 5G networks. Through connected devices, humans can link everyday objects—such as kitchen appliances, vehicles, heaters, and baby monitors—to the internet, facilitating seamless interaction among individuals, processes, and materials. Affordable computing, online storage, and extensive data processing capabilities empower physical artifacts to autonomously communicate and gather data with minimal human intervention [2]. This interconnected reality allows electronic devices to record, track, and optimize relationships among interconnected entities, thereby intertwining and synergistically collaborating the virtual and physical worlds.

Figure 5.1 depicts a variety of IoT uses, including home automation, communication, mobility, and monitoring of safety, a few of which are security in IoT.

Given that a solitary menace or assault could disrupt the network infrastructure or, even worse, grant a hacktivist complete control of the structure, cybersecurity assumes a crucial role in smart systems [3]. The IoT harbors highly confidential information, particularly in domains such as government or defense. Cyber attackers could pilfer data or inflict physical harm on the entire network if they exploit a frail access point or vulnerable device within the industrial devices. The information data can be secured as the organizations should routinely update software and "repair" devices to correct security vulnerabilities to prevent attacks

*Current and Future Cellular Systems: Technologies, Applications, and Challenges*, First Edition.
Edited by Garima Chopra, Suhaib Ahmed, and Shalli Rani.

**Table 5.1** Existing literature of detection of DDoS attacks in IoT.

| Author name | Paper name | Main objective | Pros | Cons | Future scope |
|---|---|---|---|---|---|
| Jesus Galeano Brajones, Javier Carmona Murillo, Juan Valenzuela-Valdes and Francisco Luna-Valero | Detecting and Mitigating DoS DDoS Attacks in IoT-Based Stateful SDN | This article describes a stateful SDN solution for identifying and preventing DDoS and Denial of Service assaults against connected devices. The platform is centred on an open-state, contemporary OpenFlow application. It makes use of in-switch abilities that have been demonstrated to be a viable tool for gathering statistics and prevents trying to send messages to the console. Following our findings, we formed a prototype solution that runs on the edge along with the Ryu SDN commands and identifies DoS and DDoS assaults on metrics of entropy. | In this paper, we have created software for the Ryu SDN controller(SDN) that uses probability values to recognize DDoS and DoS attacks. We have tested the user's performance in three cases to determine the impact of this measure under varying situations. | When the screen is large, the user's current state is damaged that leads, which suggests that the devices halt responding in response to commands from a control for state changes. As a result, a thorough investigation will be conducted to determine whether the issue is related to the switch's architecture. Furthermore, we can try featuring extra severe issues beyond Mininet as well as using other plans in the protocol SDN arts, testing with initial results techniques on the dynamic world with 100 s of hosts able to generate data traffic to test the software's efficiency in this aspect. | Future work will expand on this research to categorize other kinds of DDoS attacks which also compromisevarious analytical statistics that could aid in the prediction. Furthermore, intelligent mechanisms, like Machine Learning technologies that enable independent configuration of optimization methods, such as theta considering the dimensions in the frame which will be introduced. Lastly, we are developing a DDoS detector as a Computer Network that can be utilized as a separate device in the network. This would increase sensors' mobility and be familiar with other 5G technology key elements such as NFV. |

| | | | | | |
|---|---|---|---|---|---|
| In Lee | Internet of Things (IoT) Cybersecurity: Literature Review and IoT Cyber Risk Management | This article analyzes approaches for controlling cybersecurity risk and IoT cybersecurity technology. | In this article, we came across the seven layers of cybersecurity and further measures to ensure security which helped us to gain a lot of knowledge about it | The emphasis of this document's examination extends beyond the realm of managing risks in IoT security. The present methods for risk assessment are insufficient to handle the intricate web of linkages between the many devices. | This paper serves as a catalyst for developing an IoT cybersecurity risk management framework and quantitative methodologies for risk assessment, drawing insights from the literature review conducted herein. This study combines qualitative and quantitative methods and suggests a four-layer approach to managing IoT cyber risk. framework that acts as a crucial road map for efficient risk management |
| Murat Kuzlu, Corinne Fair, OzgurGuler | Role of Artificial Intelligence in Internet of Things (IoT) | The goal of this review paper is to thoroughly present and summarize the relevant articles in these fields. It does this by collecting data including a number of more studies and polls on the IoT, artificial intelligence, and attacks on and using AI | It describes that IoT and AI are being exploited for unlawful reasons; one form of this is the use of these technologies for nurturing flaws that may be regulated to lessen possible attacks. It'd better clarify existing threats. | Some of the models in AI are frequently still in the research and development stage or remain problematic to implement, rendering them scarce as they are not yet widely used in applications. | The models in this paper described are still in the phase of research and development, considering them uncommon and are not yet widely used in commercial applications. |

*(Continued)*

**Table 5.1** (Continued)

| Author name | Paper name | Main objective | Pros | Cons | Future scope |
|---|---|---|---|---|---|
| A. Saravanan and S. Sathya Bama | A Review on Cyber Security and the Fifth Generation Cyberattacks | This article describes a stateful SDN solution for identifying and preventing DoS and DDoS attacks in IoT networks. The system is formed by Open State, current application which takes advantage of in-switch skills and has proven to be a useful method for data capture because it avoids trying to send messages to the console. Following our findings, we formed experimental prototype that runs on edge of the Ryu SDN controller and detects DoS and DDoS attacks on entropy values | This system depends on an open -state, the present open flow application that takes advantage of in-switch skills and has proven to be a useful method for data capture because it avoids trying to send messages to the console. Following our findings, we formed a demonstration software that runs on the edge of Ryu SDN controller and detects DDoS attacks on statistics models. | When the screen is large, the user's current state is damaged that leads, indicating devices halt functioning towards the control plane's bids for state changes. As a result, a thorough investigation will be conducted to determine whether the issue is related to the switch's architecture or the SDN application. Furthermore, we will try with more severe issues outside of Mininet as well as using other plans in the protocol SDN arts, testing with initial results techniques on the dynamic world with 100 s of hosts able to generate data traffic to test the software's efficiency in this aspect. | Future work will expand on this research to categorize other kinds of DDoS attacks to include various analytical statistics that could aid in the prediction. Furthermore, intelligent mechanisms, such as Machine Learning technologies that allow self-configuration of optimization methods, such as theta and the window size, will be introduced. Lastly, we are developing a DDoS detector as a Computer Network that can be utilized as a separate device in the network. |

| | | | | | | |
|---|---|---|---|---|---|---|
| Waqas Ahmad, Aamir Rasool, Abdul Rehman Javed, Thar Baker, and Zunera Jalil | MDPI | Cybersecurity in IoT-based Cloud Computing: A Comprehensive Survey | In this paper, we explored significant risks, installation strategies, and service systems. Then, we divided cloud security concerns into 4 groups and went over the issues in both. | The scalability that cloud computing offers allows for the diffusion of assets and information across multiple areas and availability out of a range of commercial settings. The way services are being used, managed, and exchanged for commercial processes evolved due to virtualization, including media, resources, and tools. | The use of cloud computing increased a company's adaptability and sustainability, enabling it to stay cutting-edge and viable in a constantly evolving industry. However, because of several factors, they also made user data less safe and attackable. | Future studies may concentrate on data encryption challenges throughout the existing data centers and offer other digital logic ways to improve data protection. |
| Giannoutakis, K. M.; Spathoulas, G.; Filelis-Papadopoulos, Christos K.; Collen, A.; Anagnostopoulos, M.; Votis, K.; Nijdam, N. A. | Ucc. ie | A blockchain solution for enhancing cybersecurity defense of IoT | In this paper, a modular way for enhancing IoT security in home automation via blockchain technology was developed. | The model enables code that enables user accounts and the delisting of illegal IPs, as well as trust actions in preserving all security of equipment parts that make up home automation. | As these kinds of gadgets have a minimal amount of power, they act as light devices in the system, requiring the addition of extra full units with sufficient processor speed. | This idea presents a solution free of blockchain technology that is quickly deployed for such setup of only a few needs also on home automation ports. |

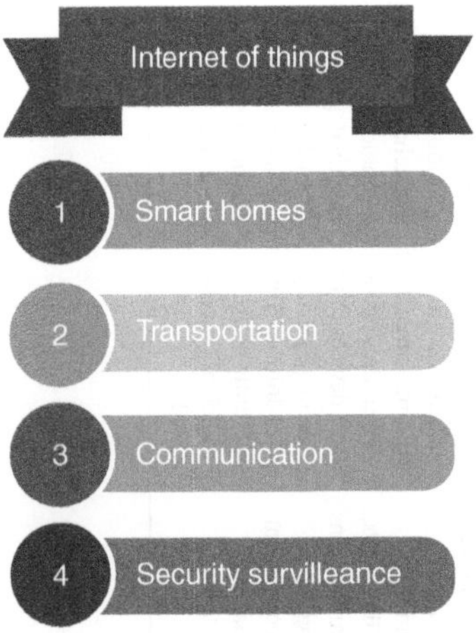


**Figure 5.1** Internet of things.

on the IoT [4]. The devices are individually protected using this technique, making the technology hard to alter. Additionally, companies must prevent using default credentialson various platforms and machines, notably within the IoT. Multifactor verification can add a layer of protection to login and password. In the end, the risk reduction strategy will help prepare the industry with secure IoT techniques. Through this paper, we can come across various practices by which we can protect our system and come to know about various attacks.

### 5.1.1 History of IoT and Edge Computing in 5G

The historical progression of IoT and edge computing delineates a narrative of technological evolution and its profound influence on reshaping our digital realm. The genesis of the Internet of Things (IoT) can be traced to the late twentieth century, where early concepts centered on the notion of interconnected devices capable of seamless communication [5, 6]. This concept gained momentum in the early 2000s with the widespread adoption of wireless technologies like Wi-Fi and Bluetooth, facilitating device communication without physical tethering. As IoT applications burgeoned across diverse sectors in the 2010s, the imperative for efficient data processing and analysis became increasingly evident. This catalyzed the advent of edge computing, initially conceived as a response to the escalating volume of data generated by IoT devices. Edge computing, though not initially labeled as such, entailed processing data closer to its origin, at the network's periphery, to mitigate latency and bandwidth consumption.

The fusion of IoT and edge computing has gathered momentum in recent years, propelled by the quest for expedited decision-making, enhanced reliability, and reduced reliance on centralized data centers. The prominence of edge computing architecture has surged, furnishing localized processing capabilities and facilitating real-time data analysis at or near the point of origin [7]. This convergence has ignited innovation across myriad domains, including autonomous vehicles, smart cities, industrial automation, and healthcare, reshaping our interaction with technology and paving the way for a more interconnected and intelligent future.

## 5.2 Literature Review

The proliferation of IoT devices has resulted in unprecedented accessibility and simplicity, but it additionally presents significant cybersecurity challenges. The goal of this overview of the literature is to present a thorough analysis of the state of cybersecurity in relation to the Internet of Things, taking into account current issues, solutions, and potential paths for protecting IoT ecosystems. The Table 5.1 highlights the literature survey.

## 5.3 Applications in IoT and Edge Computing

Edge computing harnesses everyday data and interconnected devices worldwide, providing secure features to authorized users. Individuals access global support through the IoT security measures system. The efficacy of lifespan management, firm sharing, and enhanced associations and connectivity increases with the level of techniques and solutions. Cables, sensors, and management devices underpin the economy. However, this approach's shortcomings are compounded by the expanding scope and range of security risks [8]. Cybersecurity threats to networks and infrastructures have escalated across various sectors, such as healthcare, urban planning and architecture, mobility, and emerging platforms. We will explore a range of IoT applications, including automated farming, connected vehicles, and smart residences.

The utilization of IoT in cybersecurity spans various areas, presenting innovative approaches to bolster threat detection, respond to incidents, and enhance overall resilience in digital settings. Below are several primary applications:

1) **Enhancing Network Monitoring and Intrusion Detection:** IoT devices can be employed for real-time monitoring of network traffic, ensuring continuous observation of the digital environment [9]. By employing machine learning algorithms to scrutinize network data, anomalies suggestive of cyberattacks can be swiftly identified, enabling proactive measures for threat mitigation.
2) **Strengthening Endpoint Security:** Equipped with sensors and actuators, IoT devices serve as intelligent endpoints to fortify network access. These

devices enforce access control policies, detect unauthorized activities, and autonomously respond to security incidents, thus reinforcing endpoint security posture.

3) **Facilitating Asset Management and Vulnerability Assessment:** IoT-based asset management systems track connected device inventory, pinpoint vulnerabilities, and prioritize mitigation efforts based on risk assessments. This enables proactive security measures like patch management and configuration hardening to mitigate potential threats.
4) **Integrating Physical Security with Cybersecurity:** IoT technologies integrate physical security systems, such as surveillance cameras and access control systems, with cybersecurity infrastructure to ensure comprehensive security coverage. For example, IoT-enabled access control systems enforce biometric authentication and monitor access logs in real-time to prevent unauthorized entry [10].
5) **Streamlining Security Automation and Orchestration:** Integration of IoT devices into security automation and orchestration platforms streamlines incident response workflows. Leveraging IoT sensors and actuators, security incidents are automatically detected, analyzed, and mitigated in a coordinated manner, thereby reducing response times and minimizing the impact of cyberattacks.
6) **Contributing to Threat Intelligence and Information Sharing:** IoT devices aid in gathering and sharing threat intelligence by collecting and analyzing security-related data from diverse sources. Aggregated, anonymized information is then shared with relevant stakeholders to enhance situational awareness and facilitate collaborative cybersecurity efforts.
7) **Enhancing Supply Chain Security:** IoT-enabled supply chain monitoring systems improve the security and integrity of the supply chain by tracking goods movement, verifying product authenticity, and detecting tampering or counterfeiting attempts. This mitigates risks associated with supply chain attacks and ensures the integrity of critical assets.
8) **Utilizing IoT for Security Training and Awareness:** IoT devices support interactive cybersecurity training and awareness programs, simulating realistic attack scenarios and offering hands-on learning experiences for users. By engaging employees in immersive training exercises, organizations elevate cybersecurity awareness and cultivate a culture of security-conscious behavior [11].

In conclusion, integrating IoT technologies into cybersecurity presents substantial potential for enhancing threat detection, incident response, and overall resilience against evolving cyber threats. Addressing associated challenges such as privacy concerns, data integrity, and interoperability is crucial to fully harness the benefits of IoT in cybersecurity applications.

## 5.4 Cybersecurity Management System for IoT Environments

IoT connects numerous innovative devices to a secure network. IoT safety can be a way of enhancing and merging every change linked with IoT to secure the whole safety of our planet.

### 5.4.1 Security Layers

IoT is a global web that connects goods through technology such as sensors, actuators, and handsets [12]. Due to its insufficient treatment, energy, and cache capacities, IoT is liable to several types of assaults at every level, the utmost prevalent of which is the service disruption assaults (Figure 5.2).

1) **Sensing Tier**
   The tier consists of data cables and sensors capable of detecting, gathering, processing, and transmitting information to the whole system. Four main cybersecurity problems exist in this layer: Cellular strong signals, sensor networking in IoT devices, and the development of IoT layout, unity, analysis of data, copies and outline. This layer uses three commonly used techniques to secure the IoT network:

   (1) tool for data omission
   (2) connectivity type of power
   (3) Sensor safeguards

   The goal of significant crimes and attacks, such as spoofing, approaching, session, hijacking, data manipulation, and others, is the security of the user's devices.
2) **Networking Tier**
   The tier serves as designed to protect the fundamental connectivity framework. An increasing number of communal networks, such as the Interweb, have

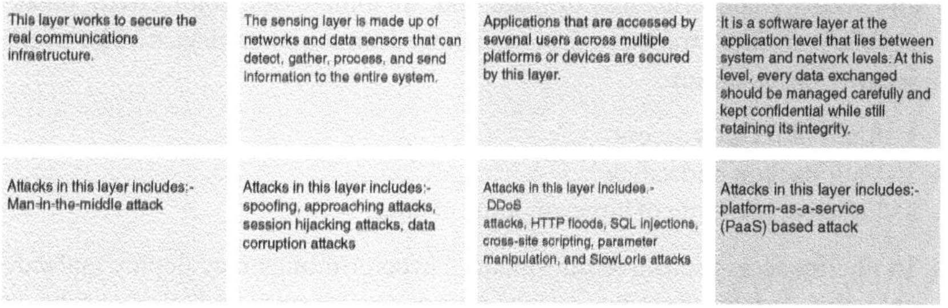


**Figure 5.2** Stratification in cybersecurity: exploring multilayered defenses.

experienced a surge in the adoption of defensive strategies at the network level for transmitting data. This is because such measures can safeguard open environments and applications without necessitating alterations to them. The layer primarily facilitates the exchange of information from a multitude of IoT gadgets across the cellular networks [13]. The primary cybersecurity concerns within this tier involve:

(1) Secrecy
(2) Data privacy
(3) Interoperability

Specifically, the networking tier is notably susceptible to a monster midway breach. The middleman attack belongs to the class of virtual assault wherein perpetrators eavesdrop on or impersonate legitimate participants to covertly monitor a recurrent conversation or resources exchange. Internet Protocol Security (IPsec) is generally embraced architecture engineered to fortify assurance within web stratum. Here are several fundamental safeguards provided by means of IPsec:

- **Secrecy**

This facilitates encrypted messaging among interacting agents while thwarting surveillance through external representatives.

- **Ensuring data reliability and authenticating**

The matter affirms that a received packet indeed originated from the institution designated in the capacity of the originator among the information packet as well as ensures wherein packet remains intact and unaltered by unauthorized entities.

- **Key management**

It facilitates the secure exchange of authentication tokens and provides protection against particular security violations, including reiteration assaults.

3) **Application Tier**

Applications that are used via numerous people across multiple channelsor devices are shielded by the tier. It is the nearest layer to end-users which makes simple with regard to cybercriminals toward target when it causes functionality and security concerns, loss of data, and, in some cases, connectivity breakdowns. Bad data is transferred and exchanged among smart devices. The layer's security problems are:

1. Right of records access
2. Data Backup and safety
3. Effective handling of large amounts of data

To illicitly access sensitive information, cybercriminals may deploy malware onto the system through trojans, viruses, or worms, and also engage in fraudulent schemes to pilfer authentication credentials. **DDoS assaults, HTTP inundations, SQL infiltrations, cross-domain scripting, variable tampering, and SlowHTTP breaches are a few couple of instances.** Many businesses employ a range of protective measures in the application tier, such as web application firewall (WAF) solution, secure web gateway solutions, additional safeguards, to counteract these risks and beyond.

4) **Intermediate Stratum**
   The Service-Oriented Architecture (SOA) principle serves as the foundation for the intermediate tier. It constitutes a software stratum positioned amidst the network infrastructure and the operating environment. Genuineness, coherence, and secrecy of all data exchanged at this level require management and supervision. Using the IoT architecture, intelligent middleware may establish fluid systems for the tangible realm within the cyber/simulated sphere via merging in-comprehensive data. Internal adversaries capable of altering and obtaining insights within the system will undertake hostile insider attacks. Underlying intrusions include PaaS attacks. Coders' goal aims to maintain the connected devices and reduce levels safe.

## 5.5 Current Cyber Security Strategies in IoT

As the Internet of Things (IoT) continues to expand, so do the associated cybersecurity risks. In this introduction, we'll delve into the current landscape of cybersecurity strategies tailored specifically for IoT environments. From threat detection to incident response, understanding these strategies is essential for safeguarding IoT ecosystems against evolving cyber threats [14].

IoT devices, ranging from smart home appliances to industrial sensors, have become integral components of modern infrastructures. However, their widespread adoption has also introduced new vulnerabilities, making them lucrative targets for cybercriminals. Traditional cybersecurity approaches often fall short in addressing the unique challenges posed by IoT, necessitating the development of specialized strategies.

1) **Encryption**
   Securing data while it's dormant and while it's in motion is critical for ensuring the security and authenticity of confidential data preserved or communicated by connected devices. Some optimal approaches for encrypting data while dormant and during transmission are:
   - Encryption of stationary data: Encrypt the data residing within smart devices, including user authentication details, device settings, and critical operational information. This approach assists to safeguard data stored within unwanted access to the event due to security incident or unauthorized

removal. Data at rest ciphering should be carried out using industry-standard encryption techniques such as the Advanced Encryption Standard (AES).

- Secure Encryption Algorithm Selection: Opt for established and widely vetted cryptographic algorithms like AES, RSA, or Elliptic Curve Cryptography (ECC), which have undergone rigorous evaluation and testing within the security experts. Refrain from utilizing closed-source or uncertified encryption techniques, as they could potentially harbor vulnerabilities or weaknesses exploitable by attackers.
- Encrypted Key Establishment: Utilize robust key establishment protocols, like Diffie-Hellman or Elliptic Curve Diffie-Hellman (ECDH), to generate secured encryption between IoT devices concurrently protecting them from potential eavesdroppers.
- Conduct regular resilience assessments on encryption solutions, such as ethical hacking, risk evaluations, in addition to scrutinizing the code for, to identify and rectify any shortcomings or vulnerabilities that could arrise.

2) **Blockchain**

Blockchain technology, along with similar technologies, does not provide a panacea for cybersecurity problems. They only endorse contemporary endeavors aimed at ensuring secure networks, communications, and data, if at all. Blockchain preserves permanent information through hashing and cryptographic transformation, a lot of the present cybersecurity solutions employ closely related technologies. Most security systems in place today rely on a single, reliable source to store encrypted data or validate information. Therefore, the system becomes vulnerable.

Abundant hostile participants may focus their endeavors on one particular aim to initiate service disruption assaults, insert deleterious material, or get confidential information through theft or blackmail. Since genuine blockchains operate in a decentralized manner and refrain from the governing body or confidence of an individual group or a member within the network, they are superior to existing security mechanisms. Since every every node, or participant, maintains a full archive of all past data that is Accessible, the system operates without necessitating trust because new assets can solely append to the sequence of previously available data via reaching a predominant agreement.

## 5.6 IoT Cybersecurity's Role in Reshaping Machine Learning

In Figure 5.3, supervised learning and unsupervised learning are two types in which machine learning can be divided. With respect to supervised learning, labeled data corresponds to clearly distinguished as either suspicious or genuine

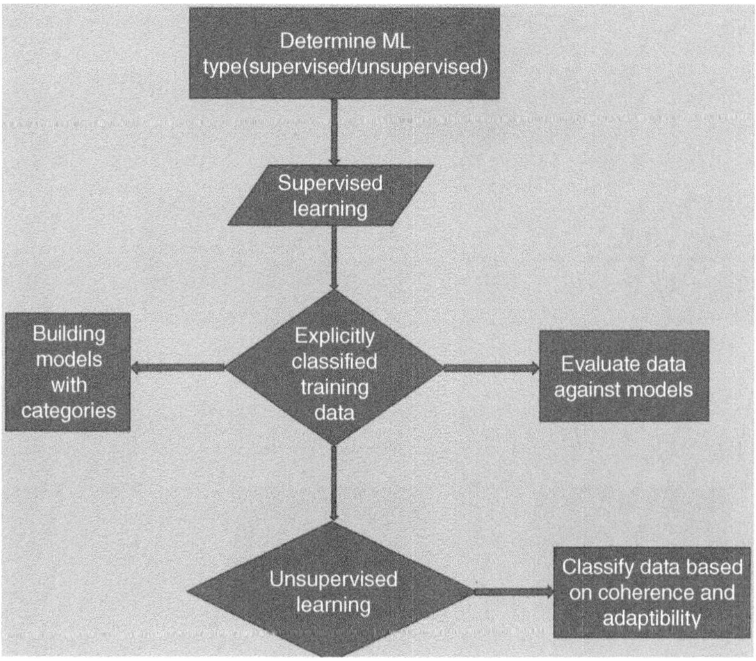


**Figure 5.3** Machine learning security architecture.

prior to being input to synchronize with a computer to generate featuring "categories" that provide benchmarks for evaluating the analyzed data. Unsupervised learning divides data based on the unity of the information within a class as well as the adaptability of the data among categories, in contradistinction to supervised learning, which employs training dataset and hand labelling. Naive Bayes stands out as a renowned machine learning method in cybersecurity that uses the Bayesian theorem to categorise data, assuming that odd behaviours arise from distinct events rather than a single attack. It's a supervised learning algorithm that, post-training and class prediction, evaluates the probability of each action being anomalous.

### 5.6.1 Role of IoT in Artificial Intelligence

Artificial intelligence refers to software that enables technology to learn and adapt to improve systems' perception, reasoning, and behavior as shown in Figure 5.4. It has been used much more frequently in business applications in recent years, such as decision support and customer service. To facilitate the deciphering and rapid analysis of data by machine learning (ML) solutions,

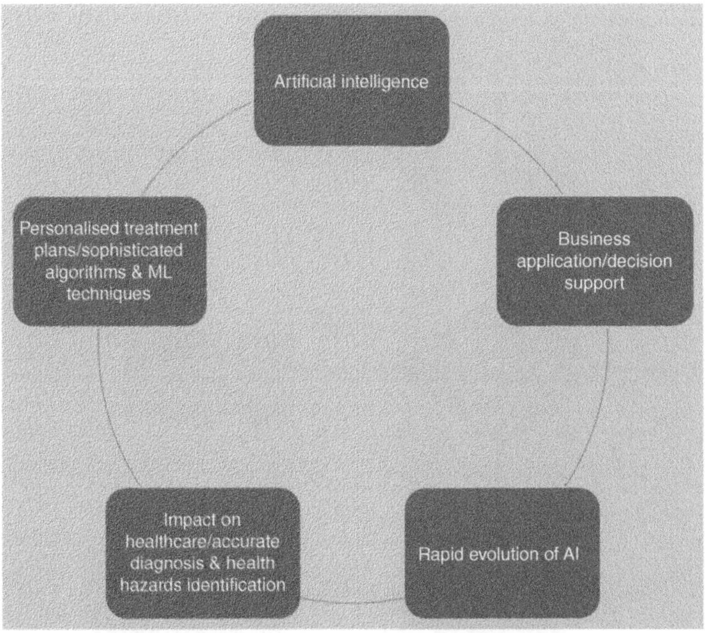


**Figure 5.4** IoT-enabled artificial intelligence ecosystem.

security experts advise standardizing the data sets that are available. The science of artificial intelligence (AI) is expanding quickly and changing many facets of our lives, including healthcare. AI can assist medical professionals with diagnosing patients more accurately, identifying possible health hazards, and creating individualized treatment plans for each patient by utilizing sophisticated algorithms and machine learning techniques.

## 5.7 Real Life Scenario

The German national railway network, Deutsche Bahn, experienced a malware attack in May 2017 known as WannaCry, resulting in the disruption of 450 of its systems, as illustrated in Figure 5.5. The attack led to the shutdown of the corporation's passenger and cargo logistics, along with additional offerings. Shortly thereafter, the federal train service belonging to both China and Russia were seen as subjected to a comparable attack, originating from North Korea. The worldwide food distribution network has undergone a significant shift with the emergence of autonomous and advanced logistics, intelligent GPS devices, and other technological advancements. These innovations have not only greatly enhanced security, quality, and efficiency but also strengthened app support.

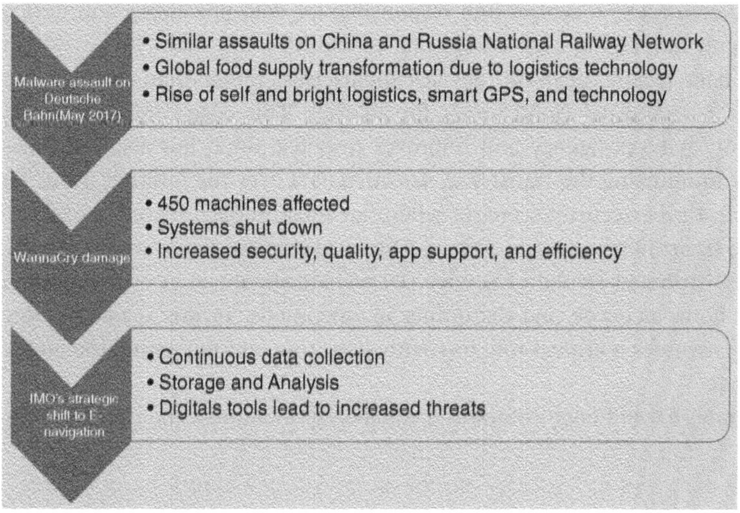


**Figure 5.5** Scenario visualization.

However, the use of digital tools has accelerated access to threats, that are usually powered by intelligent, sensor-equipped IoT technology.

This automotive industry seems grappling with the widespread nature of vulnerabilities as it confronts the threats it encounters, as evidenced by previous cybercrimes targeting shippers. This is primarily attributed to the interconnectivity with external appliances, deficiencies among numerous architects, or a widespread shortage pertaining to security experts. A significant portion in reference to the connectivity, for example through tramway systems, is commonly facilitated through WLAN or cellular connections. The rail bank's open ports are expanded, thereby reopening a wide array of vulnerabilities that can be exploited to gain control over network weaknesses. As observed, rail line sector faces a broader exposure of open ports prompted by the utilization of smart - connected automated systems or mechanized processes, or procedural operations. This is due to their handling of large volumes of data, which renders them valuable prime objectives for adversaries. A breach of the guidance system would have an impact on every facet of shipping functions, for instance, because the International Maritime Organization's (IMO) significant change in strategy enables continuous gathering, and examination of vessel and container information to document port locations, cargo details, maintenance concerns, and additional data.

**In the realm of cybersecurity, one tangible example where IoT and edge computing converge is within smart home security systems.**

Imagine a smart home outfitted with an array of IoT devices including smart cameras, door locks, motion sensors, and environmental monitors. These devices seamlessly connect to a central hub responsible for data processing and device management. Leveraging edge computing capabilities, the smart home system processes data locally, near the devices themselves, rather than transmitting all data to a centralized cloud server for analysis. This edge computing strategy not only reduces latency and improves response times but also enhances privacy by minimizing the transfer of sensitive data over the internet. Now, let us envision a cybersecurity scenario where an unauthorized intruder seeks to breach the smart home security system. The intruder may deploy a cyberattack, such as a Distributed Denial of Service (DDoS) attack, aimed at overwhelming the smart home network and disrupting its operations. In this scenario, edge computing assumes a critical role in swiftly detecting and mitigating the attack in real-time.

Here's how IoT and edge computing collaborate to address the cybersecurity threat:

1) Anomaly Detection at the Edge: Edge computing devices continuously monitor network traffic and device behavior for anomalies suggestive of a cyberattack. By analyzing data locally, in proximity to the devices, the edge computing infrastructure can promptly identify suspicious patterns or deviations from normal behavior.
2) Localized Threat Intelligence: Equipped with threat intelligence capabilities, edge computing devices can compare network traffic and device behavior against known attack signatures or behavioral patterns associated with cyber threats. This enables rapid detection and categorization of potential security incidents.
3) Immediate Response: Upon detecting a cyberattack, the edge computing infrastructure can trigger automated response actions to mitigate the threat. For instance, it may block suspicious network traffic, quarantine compromised devices, or issue alerts to notify homeowners or security personnel.
4) Reduced Reliance on Centralized Infrastructure: By processing data locally at the edge, the smart home system lessens its dependence on centralized cloud servers for cybersecurity monitoring and response. This decentralized approach enhances resilience against cyberattacks by dispersing security functions throughout the network.

In summary, this scenario demonstrates how IoT and edge computing technologies collaborate to fortify cybersecurity within real-life contexts such as smart home security systems. By harnessing edge computing capabilities for localized data processing and analysis, IoT devices can effectively detect, respond to, and mitigate cybersecurity threats in real-time, thereby safeguarding the integrity and privacy of smart home environments.

## 5.8 Conclusions

Robotic cars, unmanned aerial vehicles, intelligent grids, advanced manufacturing, and eHealth exemplify the broad spectrum of applications facilitated by the IoT. These advancements are reshaping industries, streamlining operations, and enhancing overall quality of life. Nonetheless, as various networks interconnect within IoT ecosystems, the potential security risks are magnified. In the event of security breaches within IoT systems, malicious actors seize heightened opportunities to exploit vulnerabilities, breach confidentiality, and jeopardize critical infrastructure. Absent a robust strategy to address IoT security risks, organizations encounter substantial hurdles in effectively managing and mitigating cybersecurity threats within their IoT environments. The interconnected nature of IoT devices intensifies the intricacies of cybersecurity, necessitating comprehensive measures to defend against a myriad of potential threats. From data breaches to denial-of-service attacks, the repercussions of inadequate IoT security can be severe, impacting not only organizational functions but also posing threats to public safety and privacy. To effectively confront the evolving landscape of IoT cybersecurity threats, organizations must prioritize proactive measures. This includes implementing robust authentication mechanisms, encrypting data both during transit and at rest, maintaining vigilant surveillance for suspicious activities, and promptly addressing vulnerabilities through patch management. Furthermore, collaboration among stakeholders, including industry partners, regulators, and cybersecurity experts, is imperative to establish standards, share threat intelligence, and foster a unified response to emerging threats within the IoT realm. By embracing a proactive and collaborative approach to IoT cybersecurity, organizations can fortify their defenses, mitigate risks, and fully harness the potential of IoT technologies while safeguarding against malicious actors and upholding the integrity and resilience of IoT ecosystems.

## References


1 Lee, I. (2020). Internet of things (IoT) cybersecurity: literature review and IoT cyber risk management. *Future Internet* 12 (9): 157.

2 Lu, Y. and Da Xu, L. (2018). Internet of things (IoT) cybersecurity research: a review of current research topics. *IEEE Internet of Things Journal* 6 (2): 2103–2115.

3 Ahmad, W., Rasool, A., Javed, A.R. et al. (2021). Cyber security in IoT-based cloud computing: a comprehensive survey. *Electronics* 11 (1): 16.

4 Galeano-Brajones, J., Carmona-Murillo, J., Valenzuela-Valdés, J.F., and Luna-Valero, F. (2020). Detection and mitigation of dos and DDoS attacks in IoT-based stateful sdn: an experimental approach. *Sensors* 20 (3): 816.

5 Kuzlu, M., Fair, C., and Guler, O. (2021). Role of artificial intelligence in the internet of things (IoT) cybersecurity. *Discover Internet of Things* 1 (1): 1–4.

6 Saravanan, A. and Bama, S.S. (2019). A review on cyber security and the fifth generation cyberattacks. *Oriental Journal of Computer Science and Technology* 12 (2): 50–56.

7 Giannoutakis, K.M., Spathoulas, G., Filelis-Papadopoulos, C.K. et al. (2020). A blockchain solution for enhancing cybersecurity defense of IoT. *2020 IEEE International Conference on Blockchain (Blockchain)*, (2 November 2020), pp. 490–495, IEEE.

8 Taylor, P.J., Dargahi, T., Dehghantanha, A., Parizi, R.M. and Choo, K.K.R., (2020). A systematic literature review of blockchain cyber security. *Digital Communications and Networks* 6 (2): pp. 147–156.

9 Mughal, A.A. (2019). Cybersecurity hygiene in the era of internet of things (IoT): best practices and challenges. *Applied Research in Artificial Intelligence and Cloud Computing* 2 (1), pp. 1–31.

10 Torresen, J. (2018). A review of future and ethical perspectives of robotics and AI. *Frontiers in Robotics and AI* 4, p. 75.

11 King, M.R. (2023). The future of AI in medicine: a perspective from a Chatbot. *Annals of Biomedical Engineering* 51 (2), pp. 291–295.

12 Kuzlu, M., Fair, C. and Guler, O. (2021). Role of artificial intelligence in the internet of things (IoT) cybersecurity. *Discover Internet of things* 1 (1), p. 7.

13 Badotra, S. and Panda, S.N. (2021). SNORT based early DDoS detection system using Opendaylight and open networking operating system in software defined networking. *Cluster Computing* 24 (1): 501–513.

14 Verma, K., Bhardwaj, S., Arya, R. et al. (2019). Latest tools for data mining and machine learning. *International Journal of Innovative Technology and Exploring Engineering* 8 (9S): 18–23.

# 6

# Autonomous Systems for 5G Networks: A Comprehensive Analysis of Features Toward Generalization and Adaptability

*Durga Shankar Baggam*[1,2*] *and Shalli Rani*[2*]

[1]*Department of CSE, Gandhi Engineering College, Gandhi Vihar, Khurda, Odisha, India*
[2]*Chitkara University Institute of Engineering and Technology, Chitkara University, Rajpura, Punjab, India*

## 6.1 Introduction

Autonomous is derived from the concept of automation that was presented during the Arab and Greek periods around 1200 AD. Automation in the industry has resulted from the development of the conveyor belt; as a result, an industrial revolution has begun with numerous contributions from researchers and is being extended to numerous fields. Ctesibius designed a float regulator in 270 BC using an early feedback-controlled mechanism, which serves as a reminder of some important milestones. Based on a feedback control system, the Persian brothers Banu Musa created a number of automatic controls in 850 AD. In 1624, automatic control systems were developed to design steam engines, boilers, and furnaces during the industrial revolution in Western Europe. Pressure regulators were developed by Cornelius D in the year 1681, and speed controllers for steam engines were developed in the year 1700. In 1745, windmills were automated to produce power, and Joseph, M. J. introduced punch cards in 1800. Later in the 19th century, there was an immense commitment toward automation in various industries. In 1920, electric amplifiers used in the telephone industry to reduce noise were introduced; in 1940, the discontinuous automatic control theory was presented. In 1958, programmable logic controllers entered the field of industrial control systems. As a result, rapid advancements began in 1970, and hardware began to become less expensive. There are applications in various fields, including manufacturing automation, agricultural production, power generation, the retail and food industries, logistics, construction, mining, video surveillance,

* These authors contributed equally to this work.

*Current and Future Cellular Systems: Technologies, Applications, and Challenges*, First Edition.
Edited by Garima Chopra, Suhaib Ahmed, and Shalli Rani.

**Table 6.1** Timeline of autonomous vehicles development since 1920.

| Year | Event |
|---|---|
| 1920–1930 | Radio-controlled cars introduced as autonomous vehicles. |
| 1980–1990 | DARPA began developing autonomous vehicles for military purposes. |
| 2009 | Google began developing its self-driving car. |
| 2012 | Nevada became the first state to issue a license for autonomous vehicles. |
| 2015 | Tesla Motors Autopilot software was released, for limited autonomous driving on highways. |
| 2016 | Uber launched a pilot program for autonomous ride-sharing in Pittsburgh, Pennsylvania. |
| 2018 | Waymo started offering self-driving taxi rides to select customers in Phoenix, Arizona. |
| 2020 | Cruise was permitted to test autonomous vehicles without human safety drivers on public roads in San Francisco. |
| 2021 | Guidelines for autonomous vehicles released by the National Highway Traffic Safety Administration. |

transportation, medical diagnosis, hospital administration, smart homes, and smart cities. The autonomous vehicle's timeline is given in Table 6.1.

Autonomous systems perceive using various sensors and actuators act to the environment. Advancements in sensor communication technology, Internet of Things (IoT) protocols, data analytics, and service and security applications serve as a base. Automation is the process of incorporating predetermined decision criteria into a machine. Automation and system development is the most demanding and is adopted in these real-world issues to make human life easier. However, Industries 4.0, autonomous vehicles, and healthcare systems feature an active collection of contributions. The most-contributed articles are on autonomous vehicles, the reason behind this is traffic management which is a major failure and road safety is a primary factor in many parts of the world. All of this necessitates technology-driven traffic management, which will eventually be in demand for autonomous vehicles. The degree of autonomy in its operations should be used to evaluate autonomy as a feature.

Based on their characteristics, autonomous systems are divided into five levels [1] and the same is illustrated in Table 6.2.

The environment is perceived by autonomous systems, which then respond with actions. They comprise planning, decision-making, communication systems, user interactions, and sensor technology. However, sensors perceive the environment to gather environmental data to plan and decide the action to take after

**Table 6.2** Autonomous systems levels based on its features.

| Level | I | II | III | IV | V |
|---|---|---|---|---|---|
| Independent task execution | 0 | 1 | 2 | 3 | 3 |
| Sensors | 0 | 1 | 2 | 3 | 3 |
| Data interaction | 0 | 1 | 2 | 3 | 3 |
| Decision making | 0 | 1 | 3 | 3 | 3 |
| Human interaction | 3 | 2 | 2 | 1 | 0 |
| Communication technology | 0 | 1 | 2 | 3 | 3 |

Scale: 0-NIL;1-LOW;2-MEDIUM;3-HIGH; .
Source: Adapted from [1]

every user interaction. Artificial intelligence and machine learning algorithms enable the decision-making process and system effectiveness. System objectives are established during planning and must be satisfied through appropriate actions. Autonomous vehicles are gaining popularity [1]; they act as a tool for efficient traffic management and provide road safety. Autonomous vehicles (AVs) make life easier and make it safe and comfortable during commuting. The development of intelligent transportation systems (ITS) with a focus on traffic safety and enhancing the efficiency of transportation and informational services for users [2–4]. Safety and fuel efficiency can be achieved by the optimal speed of a vehicle; the same can be managed by routing the vehicle through the navigation system for better speed and to manage traffic [5]. Autonomous vehicles, sensors, communication technology, and knowledge processing using machine learning are a powerful contribution to an intelligent transport system [6–8].

The rest of the paper is organized as Section 6.2: Survey Method; Section 6.3: Background and Related Works; Section 6.4: Discussions; Section 6.5: Conclusion.

## 6.2 Survey Method

The search is conducted on various latest topics "autonomous systems", "autonomous vehicles", "autonomous hospital management systems" and "autonomous systems surveys" to identify the issues. Due to its wider scope and applications, there is a significant contribution to this problem space. This study is initially conducted on Google Scholar Search and then the same is conducted on popular databases like IEEE Xplore and Scopus, the same is discussed in detail.

Google Scholar is the most accessible resource to analyze due to its high availability and openness, it is one of the primary sources for understanding the importance of the topic. A search conducted on Google Scholar using the key term "autonomous systems survey" articles in a single year (2022) and a collection of 15,400 articles are found; then the search is conducted for the first quarter of 2023, and there are 2200 articles. The number of citations in a domain highlights the demand for the topic.

In IEEE Xplore Search, 3392 journal articles were found using the search term "autonomous systems" during the single year 2022. This brings the total number of articles down to 421 after being further filtered using the search term "remotely operated vehicles". As there was a huge contribution and manual search is impossible, the search process is depicted in Figure 6.1. To discuss in detail the search protocol applied as Title: "Autonomous Systems Survey" + Source: Journal + Period: 1981 to 2023, with a total of 542 articles found. After further refinement to a single year as Title: "Autonomous Systems Survey" + Source: Journal + Period: 2022," the total articles found were 146, out of which 68 were open access articles.

The detailed list of articles topic-wise and the number of contributions is given in Table 6.3.

Scopus database accessed on 19.3.23. and search conducted using search protocol for the duration 2015–2023 and found 43 articles; protocol: TITLE-ABS-KEY (autonomous+systems+survey) AND (LIMIT-TO (DOCTYPE, "re")) AND (LIMIT-TO (LANGUAGE, "English")) AND (LIMIT-TO (SUBJAREA, "ENGI")) AND (LIMIT-TO (EXACT KEYWORD, "Autonomous Vehicles")). Some analyses projected on the Scopus domain are no of documents in a year Figure 6.2,

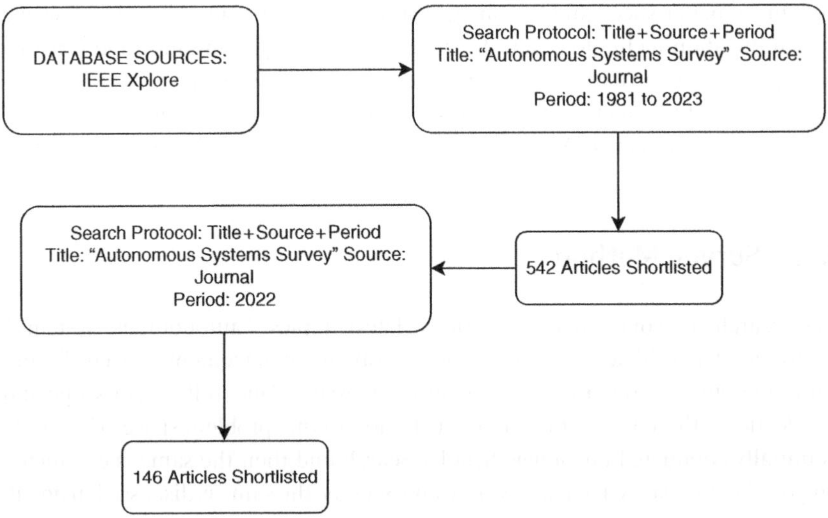


**Figure 6.1** IEEE Xplore search flow diagram.

**Table 6.3** Articles available on topic autonomous systems based on search.

| S. No. | Topic | No. of Articles |
|---|---|---|
| 1 | Autonomous aerial vehicles | 96 |
| 2 | Remotely operated vehicles | 60 |
| 3 | Road vehicles | 32 |
| 4 | Autonomous underwater vehicles | 22 |
| 5 | Underwater vehicles | 10 |
| 6 | Electric vehicles | 3 |
| 7 | Space vehicles | 3 |
| 8 | Marine vehicles | 2 |
| 9 | Unmanned surface vehicles | 2 |
| 10 | Vehicles | 2 |

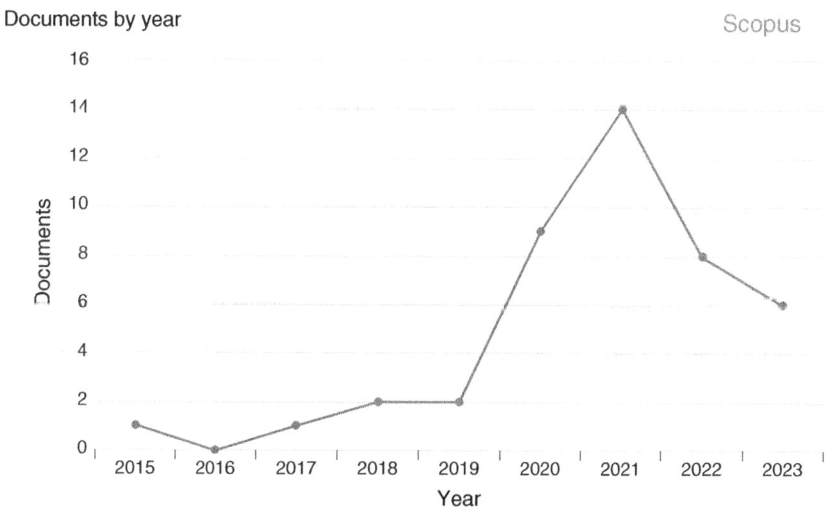


**Figure 6.2** Documents by year. Source: Scopus/With permission of Elsevier.

documents by the country Figure 6.3, documents by Subject Figure 6.4, documents by author, documents by affiliation, and documents by source. It was understood that in the initial phase, that is, 2015–2019 the contribution was very low, later phase, that is, 2020 to the current phase the contribution is higher than 400% and also raising the bar showing the demand for study in the domain.

In addition to this, a bibliometric analysis using biblioshiny R-tool for quantitative research [9] is conducted by the authors, which helps to analyze the

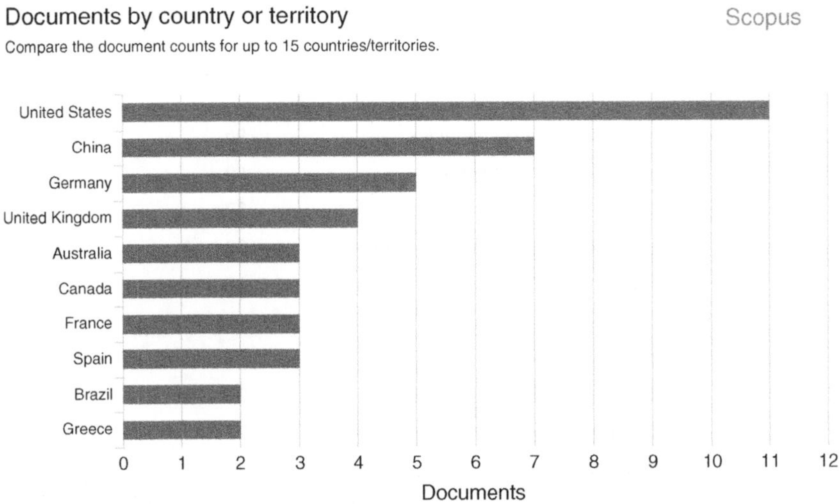


**Figure 6.3** Documents by country. Source: Scopus/With permission of Elsevier.

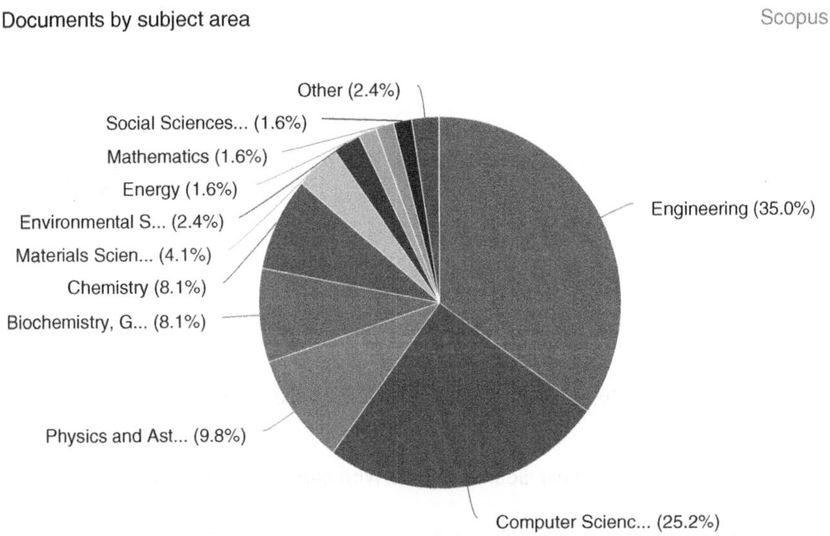


**Figure 6.4** Documents by subject area. Source: Scopus/With permission of Elsevier.

published articles and to view various metrics like Annual Growth Rate, with the same dataset obtained from Scopus and observed that the **Annual growth rate is 25.1%** for the above search protocol once again highlights the need. It also provides other parameters like productive authors, top citations, networks,

Sources, Keywords, and Thematic Maps. In addition, to the above approach, authors also conducted a detailed study of shortlisted papers to collect valuable information and any specific gap to address.

## 6.3 Background and Related Works

Despite the abundance of research on autonomous systems and autonomous cars, most of them are too specific and limited to a single area of the world. A thorough survey that is necessary to create an autonomous system is not accessible to fill the research gap. Articles are shortlisted from various databases about layered architecture, participating sensors, AI techniques used, communication protocols and technology, and device communication security issues and challenges.

According to the literature study, G.M. Amdahl first used the term architecture in 1964, and it quickly acquired recognition as a field of study in system architecture [10, 11]. Autonomous systems are basically IoT; they have no consensus on architecture and are universally agreed upon. Different researchers suggested different architectures and the popular approach is Three-Layer and Five-Layer architectures [12]. Three-layer architecture was introduced in an earlier stage with a perception layer, network layers, and application layers; Five-layer architecture contains a perception layer, transport layer, processing layer, application layer, and business layer. Recently, there has been a shift toward fog computing with the six-layered architecture and they are physical, monitoring, preprocessing, storage, security, and transport layers, a system design in which network gateways and sensors perform data processing and analytics. Power, resource, and response services are all tracked by the monitoring layer. Analytics on sensor data that is filtered is carried out by the preprocessing layer. Data replication, dissemination, and storage are the capabilities offered by the storage layer. Data integrity and privacy is assured by the security layer. On the network edge, monitoring and preprocessing are carried out before data is sent to the cloud. The concept of automation and its development toward a fully autonomous system demands independent of perceiving, planning, execution, and learning with a seamless collaboration of sensor technologies within an efficient architecture that should be extensible, evolvable, and collaborative.

According to the literature on sensors, the authors of [13–17] emphasized and elaborated in terms of sensor features and its performance that fit into a pervasive layer of autonomous systems. It can also increase the system's ability to self-adapt and self-learn [18]. Perceiving the data from the agent environment is a vital activity of a sensor, sensor communication, and its technical performance and challenges are discussed [16]. A Study of autonomous activities using deep reinforcement learning to process is discussed [19]. Possible datasets and analytic tools to design autonomous driving systems are listed with an insight into the current challenges [20].

ITS and accurate positioning systems are always in demand and need to address uncertainties to improve the behavior of autonomous vehicles articulated [21]. Wireless body networks or sensors arranged on the roadside users to reduce accidents based on behavior analysis to address various challenges like energy efficiency, self-recovery,self-reliable communication should be developed [22].

It also identifies various suitable sensors for gathering environmental data, as well as different categories of communication technologies that enable Vehicle-to-Everything (V2X) communications between entities on roads, including cars, pedestrians, etc. The advantages, disadvantages, and challenges associated with these AVs are discussed, along with how latency characteristics can be used to differentiate each category for suitable applications [23].

Understanding sensors and their fusion technology challenges and countermeasures for adversarial attacks are not covered, for a detailed study refer to [20, 24–29]. Another survey article about connected autonomous vehicles (CAVs) security [30] covers various security attacks and their countermeasures for a period of two decades, a detailed discussion on various attack models and their target components, access controls, and motives for the attack are briefly discussed. Attacks and security challenges in CAVs and their defense techniques are analyzed and found no evidence to show that the current market-ready CAV are able to defend against some novel attack models [30].

B5G-based communication in autonomous systems and its challenges have been widely discussed in recent days. In large wireless networks, operating state-space complexity is high. It is gaining popularity after embedding machine learning algorithms like deep reinforcement learning in vehicular networks, and the use cases for traffic management and control, user safety, and RSU-based infotainment. This involves decision-making by reducing human interaction to avoid uncertainty in the existing system using Deep Reinforcement Learning, which is a combination of Reinforcement Learning and Deep Learning [31]. Application-specific algorithms for data transmission are vertical handover strategy, QoS, and Switching technology and an Fuzzy Control Traffic Steering (FCTS) for better latency, throughput, delay, packetloss to avoid frequent handover [32]. In addition to this, there are numerous articles listed along with their citation count given in Table 6.4.

### 6.3.1 Autonomous System Architecture

Architecture for autonomous systems should generally be constantly evolving. Mostly system architecture is adopted to enhance autonomy, such as perception, planning, function, and information feedback. On the other hand, the logical and hierarchical links between the modules must be specified in the autonomous

**Table 6.4** Autonomous systems scopus articles and its citation count.

| Ref. no. | Title | Publication year | Cited count |
|---|---|---|---|
| Jain et al. [33] | A comprehensive review on design of autonomous robotic boat for rescue applications | 2021 | 1 |
| Magnussen et al. [34] | A survey of the inadequacies in traffic sign recognition systems for autonomous vehicles | 2020 | 7 |
| Zhang et al. [35] | Survey on deep learning-based marine object detection | 2021 | 7 |
| Badue et al. [36] | Self-driving cars: A survey | 2021 | 287 |
| Manzoor et al. [37] | 3d recognition based on sensor modalities for robotic systems: A survey | 2021 | 3 |
| Nacpil et al. [38] | Application of physiological sensors for personalization in semi-autonomous driving: A review | 2021 | 3 |
| Zhang et al. [39] | Perception and sensing for autonomous vehicles under adverse weather conditions: A survey | 2023 | 1 |
| Lopez et al. [40] | Security of emergent automotive systems: A tutorial introduction and perspectives on practice | 2019 | 12 |
| De Sousa and Drummond [41] | Metropolitan optical networks: A survey on single-layer architectures | 2023 | 0 |
| Camara et al. [42] | Pedestrian models for autonomous driving Part I: Low-level models, from sensing to tracking | 2021 | 22 |
| Li et al. [43] | Polytopic LPV approaches for intelligent automotive systems: State of the art and future challenges | 2021 | 40 |
| Laconte et al. [44] | A survey of localization methods for autonomous vehicles in highway scenarios | 2022 | 3 |
| Gomes et al. [45] | A survey on ground segmentation methods for automotive LiDAR sensors | 2023 | 0 |
| Zhang and Letaief [46] | Mobile edge intelligence and computing for the Internet of Vehicles | 2020 | 194 |
| Alaba and Ball [47] | A survey on deep-learning-based LiDAR 3D object detection for autonomous driving | 2022 | 1 |
| Khoshavi et al. [48] | Blockchain applications to improve operation and Security of transportation systems: A survey† | 2021 | 11 |
| Mohammed et al. [49] | The perception system of intelligent ground vehicles in all weather conditions: A systematic literature review | 2020 | 35 |

*(Continued)*

**Table 6.4** (Continued)

| Ref. no. | Title | Publication year | Cited count |
|---|---|---|---|
| Mohamed et al. [50] | Literature survey for autonomous vehicles: Sensor fusion, computer vision, system identification, and fault tolerance | 2018 | 19 |
| Wei et al. [51] | Survey of connected automated vehicle perception mode: from autonomy to interaction | 2019 | 22 |
| Mostaani et al. [52] | Task-oriented communication design in cyber-physical systems: A survey on theory and applications | 2022 | 0 |
| Sanchez et al. [53] | Use of UIoT for offshore surveys through autonomous vehicles | 2021 | 7 |
| Alghodhaifi and Lakshmanan [54] | Autonomous vehicle evaluation: A comprehensive survey on modeling and simulation approaches | 2021 | 10 |
| Shi et al. [55] | Algorithm and hardware implementation for visual perception system in autonomous vehicle: A survey | 2017 | 99 |
| Belwafi et al. [56] | Unmanned aerial vehicles' remote identification: A tutorial and survey | 2022 | 0 |
| Devi et al. [57] | A comprehensive survey on autonomous driving cars: A perspective view | 2020 | 14 |
| Banafaa et al. [58] | 6G mobile communication technology: Requirements, targets, applications, challenges, advantages, and opportunities | 2023 | 5 |
| Chowdhury et al. [59] | Attacks on self-driving cars and their countermeasures: A survey | 2020 | 25 |
| Dai et al. [60] | A review of 3d object detection for autonomous driving of electric vehicles | 2021 | 4 |
| Ahangar et al. [23] | A survey of autonomous vehicles: Enabling communication technologies and challenges | 2021 | 73 |
| Ridao et al. [61] | Intervention AUVs: The next challenge | 2015 | 72 |
| Wang et al. [62] | Security issues and solutions for connected and autonomous vehicles in a sustainable city: A survey | 2022 | 2 |
| Kloukiniotis et al. [63] | Countering adversarial attacks on autonomous vehicles using denoising techniques: A review | 2022 | 4 |
| Balkus et al. [64] | A survey of collaborative machine learning using 5G vehicular communications | 2022 | 11 |

**Table 6.4** (Continued)

| Ref. no. | Title | Publication year | Cited count |
|---|---|---|---|
| Rodríguez et al. [65] | A survey of machine and deep learning methods for privacy protection in the Internet of Things | 2023 | 0 |
| Schwarting et al. [66] | Planning and decision-making for autonomous vehicles | 2018 | 397 |
| Romero et al. [67] | Road curb detection: A historical survey | 2021 | 5 |
| Riedmaier et al. [68] | Survey on scenario-based safety assessment of automated vehicles | 2020 | 116 |
| Zhou et al. [69] | MMW radar-based technologies in autonomous driving: A review | 2020 | 36 |
| Malik et al. [70] | How do autonomous vehicles decide? | 2023 | 0 |
| Sulieman et al. [71] | Edge-oriented computing: A survey on research and use cases | 2022 | 8 |
| Baruah et al. [72] | Achieving resiliency and behavior assurance in autonomous navigation: An industry perspective | 2020 | 3 |
| Olakanmi [73] | SAPMS: A secure and anonymous parking management system for autonomous vehicles | 2020 | 9 |

system architecture. Autonomous system internal structures should be described in terms of the conceptual and organizational framework. The objective of architecture is to enhance the system's capacity to act autonomously on perceiving objects from the environment. The basics of architecture and its types are listed in Table 6.5.

A human action model, user state monitoring, intent detection algorithms, and effective user interfaces that promote cooperation between the user and the system are necessary to construct reliable autonomous systems. Work is being done in facial and emotional identification that would be useful in monitoring a user's status, while augmented reality devices, gestures, and voice. User interface would facilitate simple communication. These newly developing fields of artificial intelligence and object detection are building blocks. When a generalized autonomous system design complies with predetermined guidelines without making a mistake, it is the autonomous system with the ability to effectively observe and assess the environment and the system's condition to accomplish the task. A system is autonomous if it is extensible, evolves, and collaborates. One such system designed with capabilities is the autonomous waterborne transportation system that is depicted in Figure 6.5. This architecture satisfies

**Table 6.5** Comparison of architecture characteristics of automation system.

| Type | Characteristic | Deficiency |
|---|---|---|
| Hybrid design | Having a reasonably straightforward structure and being simple to deploy in addition to assuring real-time and task variety. | The system's design is extremely complex and is only workable by basic users. Moreover, system simplicity and transition redundancy are not great. |
| Cloud-based Design | In order to increase the ability of information exchange and task execution among people, complicated computer jobs are uploaded to the cloud. | Sharing information is expensive and is very reliant on the state of the network. |
| Behavior-based design | A parallel framework that responds quickly and is both strong and flexible. | It is challenging to design, and coordinating across modules is challenging as well. |
| Knowledge-based design | Each functional module is presented in a clear hierarchy that makes it easy to develop and execute them. | Lack of dependability and action was taken in real-time. |

Source: Comparison table from [18]/MDPI/CC BY 4.0.

interconnect capabilities to enable autonomously completing tasks such as control and service delivery through the object layer to process the scene, the cyberspace layer with new technology like argument reality and virtual reality to create digital twins, and the cognition layer with machine learning capability and application layer. Here a digital repository of physical entities is designed and configured to form a collaborative structure, and then they are organized and optimized with cognitive knowledge at the higher layer. The required features are described below.

*Extensibility*: It should be capable of E2E functionality with a well-to-do environment recognizing features, as well as the mental capacity to judge, plan, and act accordingly. Yet, the system should be flexible in its organization, expansion, and support of human–team mate interaction. Preferably, architecture should apply scenarios and be engineering focused. Instead, it might adopt an interactive nonfunctional strategy offered by systems with rules based on artificial neural networks, etc. Despite their structure, common architectures are promising; when agents are well connected, logical and cognitive strategies are employed. The capacity of an architecture to close conceptual and functional gaps between various agents operating in the autonomous system will be a crucial gauge of its usefulness.

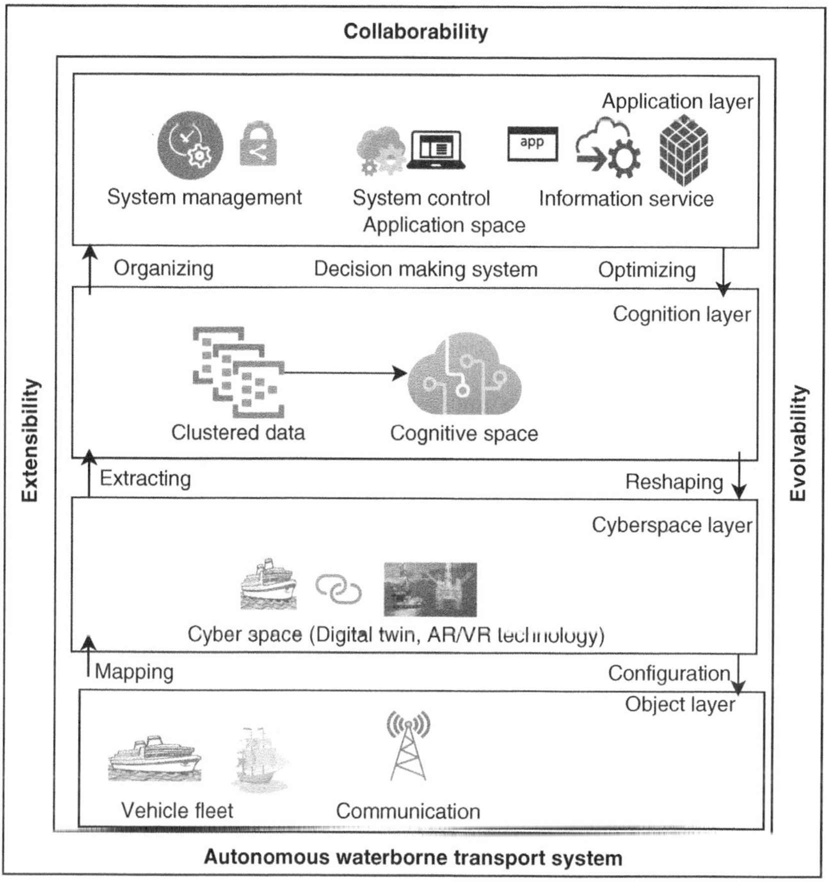


**Figure 6.5** The architecture of an autonomous waterborne transportation system.

*Evolvability*: It should facilitate technologies that can provide the necessary capabilities at different levels of autonomy while building autonomous systems architectures. These technologies enable effective knowledge-based administration, learning, and human–computer interaction. They also include basic facilities for perception, judgment, and control. From different technological perspectives, designs should be mirrored from foundational research through exploratory development and early prototyping.

*Collaborability*: The present situation demands that humans be a crucial supervisor of any autonomous system. As a result, we should give the architecture's directed design of human function a lot of thought. Before creating an architecture that more effectively supports human interactive autonomous systems, it is essential to interpret system functionality and limitations imposed on each autonomous system. The evaluation of the pertinent tradeoffs in terms of agility, detectability, observability, and transparency can be made possible by a human interactive architecture. A different task-centered design that emphasizes the objective or goal of the application rather than specific system user features is advocated for the overall architectural design scenario vignettes and application-specific use cases.

The key features are (1) pervasive cognitive mechanism design based on extensibility. (2) Self-learning and evolution mechanisms are designed based on evolvability. (3) Human–machine interaction mechanisms based on collaborability [1].

Its major integration challenges in a system are state perception [74], reliability\ [75], and human–machine interaction [76].

#### 6.3.1.1 Application Layer

This layer consists of unique consumer resources, and they are different depending upon the applications, namely autonomous vehicles, healthcare systems, smart homes, industrial devices, and smart cities. These devices perform computation on cloud infrastructure. The technical resource required to manage this is computing and communication experts. A major role at this layer is also to address challenges that occurred during collaboration and interaction with external devices.

#### 6.3.1.2 Cognitive Layer

The layer is the key to realizing autonomy, and it consists of a decision-making mechanism that acquires the cleansed data from the perception layer as input to develop a knowledge base from its inference mechanism as it evolves. This enables evolvability through knowledge management. It is used to perform analytics on detected objects to control the system. The cognitive layer infers the data through its inference mechanism to activate the entities in the system and establishes cognitive relations between object entities to generate decision-support knowledge.

#### 6.3.1.3 Perception Layer

It realizes the sensor data and its relation to knowledge space. This layer may contain some sensors, and the data received should be cleaned, which is done with the help of machine learning algorithms. It also acts as a mediator for the cognitive layer and physical layer, and it also ensures a stable and optimized communication network.

#### 6.3.1.4 Physical Layer

It describes physical objects and their characteristics involved in autonomous systems to say that traffic lights, pedestrians, vehicles, footpaths, navigation facilities, and the environment are objects of an autonomous vehicle. They are the basic entities of the entire system that realize its functionality. The physical layer is responsible for the identification of various software and hardware devices participating in the system.

### 6.3.2 Sensors

In the perception step, the environment is sensed using a variety of sensors to determine its present condition. Sensor classification of four types is presented here, and there are still many beyond the scope. The taxonomy of sensors is listed in Figure 6.6. However, RADAR, Light Detection and Ranging (LIDAR), cameras, real-time kinetic (RTK), and other sensors are the most often utilized. The recognition modules receive the data from these sensors and process it. The autonomous systems sensor components, device positioning, and localization modules collect data, which are combined and then forwarded to the following layer [23]. Review articles on sensors [13–17] are emphasized and elaborated in terms of sensor features and its performance that fit into a pervasive layer of autonomous systems. Sensors mostly used are vision cameras, LIDARs, and radars for sensor calibration, fusion techniques, and object detection. A detailed report on the performance of various sensors is given in Table 6.6. In this table, sensors are classified into four types based on the domain they deploy. Type-1 is electric, biological and chemical sensors. Type-2 is electron, electro and thermal sensors; Type-3 is photo, active, and passive sensors, and the last type are analog and digital sensors. These sensors have a vital role in the environment and many sensors are available and developed to the demand in the domain.

Based on the greatest cumulative value derived from several criteria, Camera, Radar, and LIDAR are rated as first, third, and second in comparing sensor performance under specific predetermined conditions. Yet, in autonomous systems, appropriate sensors must be used depending on the issue statement or the condition in the real world. Intrinsic, extrinsic, and temporal calibration are the three main categories of sensors. They are (1) feature-based fusion, which combines the features of different sensors into a single feature vector. (2) Model-based fusion, which uses models from multiple sources and fuses them together for better accuracy. (3) Decision level fusion, where decisions are made based on data collected by all available sensors to make more accurate predictions or detections, is also discussed, along with a review of the most cutting-edge methods for object detection [16]. However, manufacturing reliable and robust sensors is a challenging task. Managing faulty sensors and unreliable data prediction to cleanse the data need machine learning algorithms during perception.

It concludes by highlighting challenges in the sensor fusion field and proposing possible future research directions.

### 6.3.3 Artificial Intelligence Techniques

AI processes data collected through cameras and sensors, object and scene detection from recorded videos to maintain a log. The sensors used are listed in

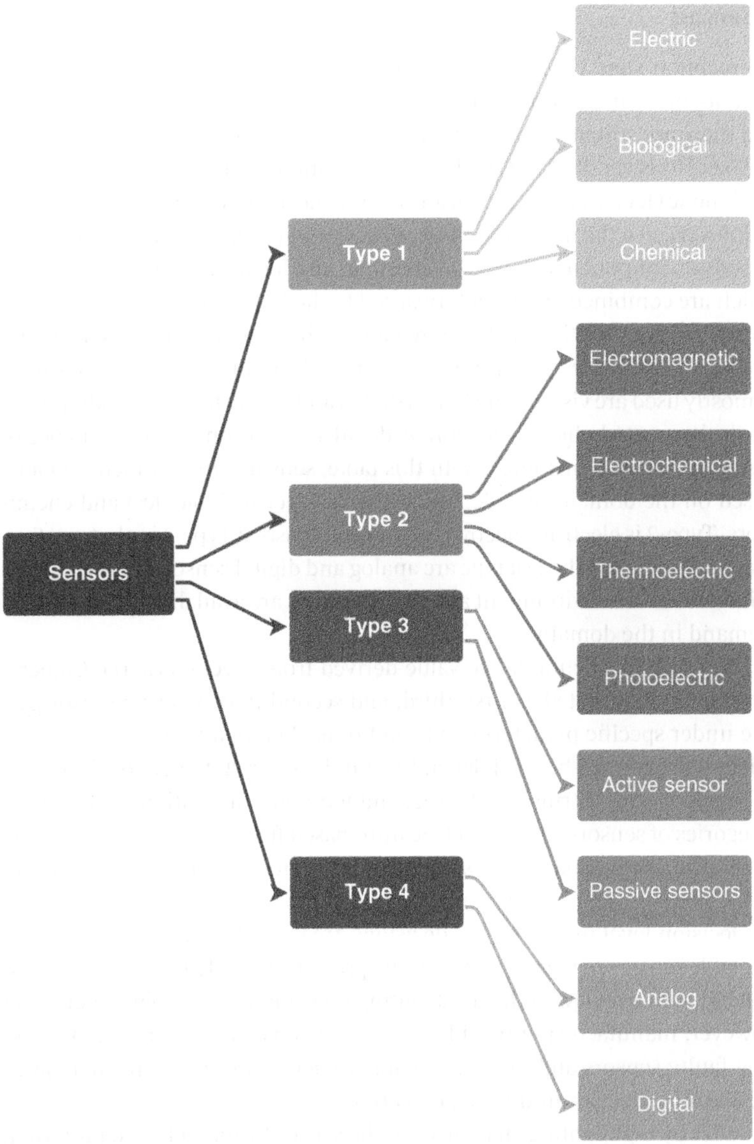


**Figure 6.6** Sensor taxonomy diagram.

**Table 6.6** Sensor performance comparison under predefined conditions.

| Device factors | Camera | LiDAR | RADAR |
|---|---|---|---|
| Range | 2 | 2 | 3 |
| Resolution | 3 | 2 | 1 |
| Distance accuracy | 2 | 3 | 3 |
| Velocity | 2 | 1 | 3 |
| Color perception | 3 | 1 | 1 |
| Object detection | 1 | 3 | 3 |
| Object classification | 3 | 1 | 1 |
| Lane detection | 3 | 1 | 1 |
| Obstacle edge detection | 3 | 3 | 1 |
| Illuminations conditions | 1 | 3 | 3 |
| Weather conditions | 1 | 2 | 3 |

Performance level: 3 – Satisfactory, 2 – Unsatisfactory, 1 – Poor.

Figure 6.6; they perceive environmental data in some constant intervals of time, and their reliability is a major challenge. There are many reviews conducted on artificial intelligence and machine learning techniques used to enhance object detection and a review article [19] on preprocessing techniques suitable for autonomous vehicles (AVs) for the purpose of reducing road accidents and human injuries. AVs should be capable of the following features: image processing, feature extraction, object detection, estimating an accurate position of objects in the scene, motion control, path planning, and data classification. Existing known techniques used in various articles are convolutional neural networks (CNNs) for 2D image processing using multiple hidden layers. Region-based CNN is a dual-stage detection algorithm used for path planning, pedestrian detection, object detection, and signal detection [77]. However, Fast Region-based CNN and also Faster Region-based CNN with optimal runtime performance are available. SLAM (simultaneous localization and mapping), known as synchronized localization mapping, perceives and locates the position of an object which helps for object detection like pedestrian detection and motion control [78]. YOLO (You Only Look Once) is an alternate way of object detection based on CNN that works efficiently in a real-time scenario and is easy to deploy [79]. However, due to the limitations of machine learning algorithms and their need for human intervention, there is a demand for deep learning techniques to mimic the human brain. A taxonomy of deep learning and reinforcement learning techniques gives solutions to four challenges in autonomous driving. The most popularly used datasets are KITTI and Cityscapes, which are used to evaluate autonomous driving algorithms. Intended researchers for an experimental setup may study a detailed research paper on various available 15 popular datasets in detail [80]. On the other hand, some researchers use simulation environments like CARLA or AirSim to generate synthetic training data through cameras arranged in the vehicles and have highlighted possible future research directions like evaluating the effect of nighttime driving and snow driving conditions for fully autonomous vehicles. Video object detection accuracy is 90% in [81, 82]. However, higher accuracy is expected to improve the efficiency of autonomous systems.

### 6.3.4 Intelligent Transport System (ITS)

Modern wireless, electronic, and automated technologies are all ITS components. These technologies have the potential to integrate infrastructure, users, and vehicles as a whole. Computerized and in-vehicle advancements incorporate accuracy docking for transports, automated routing, and impact evasion frameworks. Improving air quality, reducing dependence on foreign oil, optimizing trips, reducing unnecessary miles traveled, increasing other mode use, and reducing time spent in traffic are just a few of the many benefits of ITS technologies. ITS design collects autonomous vehicle environment data using sensors and enables vehicular communication technologies to improve travel safety and security to avoid human errors [23]. Sensors deployed in an environment and different categories of communication technologies enable Vehicle to Everything (V2X) communications between entities on roads which include cars, pedestrians, traffic signals, and any other vehicles. The experiments conducted are used to collect sensors and communication technology data to identify the main advantages, disadvantages, and challenges associated with these AVs, along with how latency characteristics can be used to differentiate each category for suitable applications. The ITS-G5 service alerts are chosen after taking into account things like application-specific requirements, reliability, and the cost of implementation, among other things. Typical channel bandwidth of 10 MHz is used in field measurements and is appropriate for all safety-critical services shared via 5.9 GHz spectrum [83]. The location, dependability, and robustness of the ITS services are all improved, which increases road traffic safety. During vehicle driving, the parametric evaluation enhances situational awareness as well as user-specific safety-related operational aspects. Different on-road warning alerts are generated by the ITS agent for use in real-world measurement and execution. It has also been shown that the correspondence between RSU and vehicles accomplishes most of the framework necessities for the age of ITS administration alerts.

### 6.3.5 B5G-Based Vehicular Telecommunication

In contrast to third-generation (3G) and fourth-generation (4G-LTE), the fifth-generation (5G) network plays a significant strain on mobility management functions due to the wide variety of numerous use-cases. The future 5G and beyond B5G mobile networks are designed to exploit signals through a variety of access technologies for various services are part of mobile service globalization [84]. At the same time, the mobile user base and machine intelligence are increasing exponentially. Heavy-bandwidth multimedia sources, multimedia technology like augmented reality (AR) and virtual reality (VR), traffic management from a large cluster of sensors, also known as the Internet of Things (IoT), interference management, routing, and other similar applications and services are among the applications and services offered to devices. To understand this scenario on vehicular networks, a comprehensive review was conducted to integrate reinforcement and deep reinforcement learning algorithms used to effectively solve decision-making problems for vehicular network management [31]. The review on technical aspects of automated driving systems, such as localization, mapping, perception planning, and human–machine interfaces, are compared with various algorithms implemented to its potential use and its challenges [20]. In addition, an experimental study on real-world driving data was collected from various public databases, simulation platforms, and autonomous vehicle fleets. It also highlights some unresolved challenges which further required the study of the development of B5G vehicular networks. The data used by the authors of the article are entities such as vehicles, roadside units (RSUs), and base stations along with traffic management parameters like speed limits, congestion levels, etc., and other safety features like road safety, lane detection systems, collision avoidance algorithms, infotainment services like multimedia streaming applications, and other communication protocols related to B5G networks and also outlines some open issues which need further study for the development of B5G vehicular networks. It enables autonomous decisions to access spectrum, data transmission rate and power control, etc., which helps maximize network performance while minimizing energy consumption.

As a result, the upcoming B5G network is regarded as a highly diverse ecosystem of technologies that covers low-power nodes (LPNs), IoT-based applications, VANETS, ad hoc networks, MANET network routing and slicing, multiple-input and multiple-output (MIMO) platforms, and platforms of ultra-dense networks (UDN). Mobility and link management is a primary concerns when moving between coverage areas. The portability conventions guarantee the conveyance of valuable information and control signals through a radio connection point connected to the ideal client. However, it is impractical to deploy a uniform network in order to provide radio-connected smart devices with seamless mobility.

As a result, in order to guarantee the effective provisioning of anticipated services, the B5G networks require modification of management protocols, ultra-reliability in the handover mechanism, and flexibility in the network topology.

## 6.4 Discussion

Designing autonomous systems should address multiple factors and various system components, including agents, physical objects and their software components, control mechanisms, sensors, and actuators. It is required to build numerous perspectives for design, including system coordination, development, structural, and behavior views due to the environment's scattered aspects. As a result, it has been acknowledged that software and system designs are crucial components of process development for managing the complexity of systems also, a wide range of nonfunctional needs are listed, and formal architecture standards are seldom taken into account; in reality, autonomous system architecture design should perform functionality as given in Figure 6.7 and the four-layer architecture in Figure 6.8 and working is illustrated in a swimlane diagram in Figure 6.9.

Functional architecture as given in Figure 6.7 demonstrates module-wise components and the services provided by each layer. The modules are sensors deployed in the environment, perception, scene detection, real-world model, behavior analysis, and finally response through actuators. A high-definition camera deployed

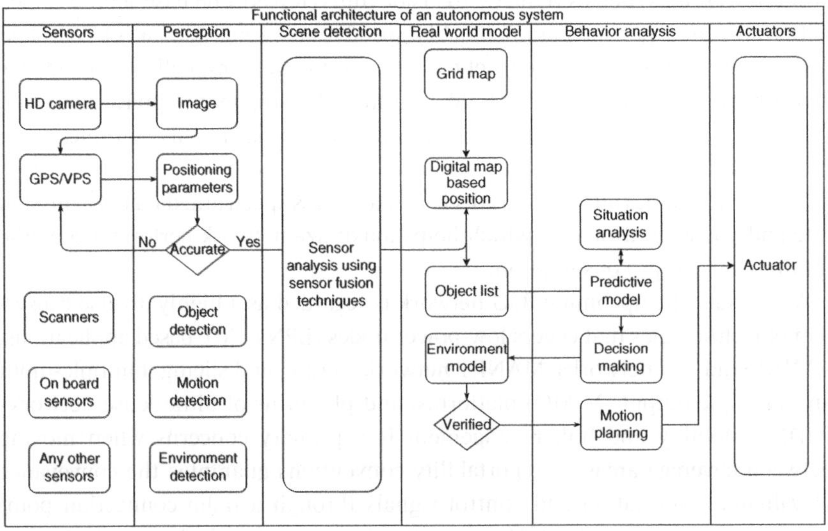


**Figure 6.7** Functional architecture of an autonomous system.

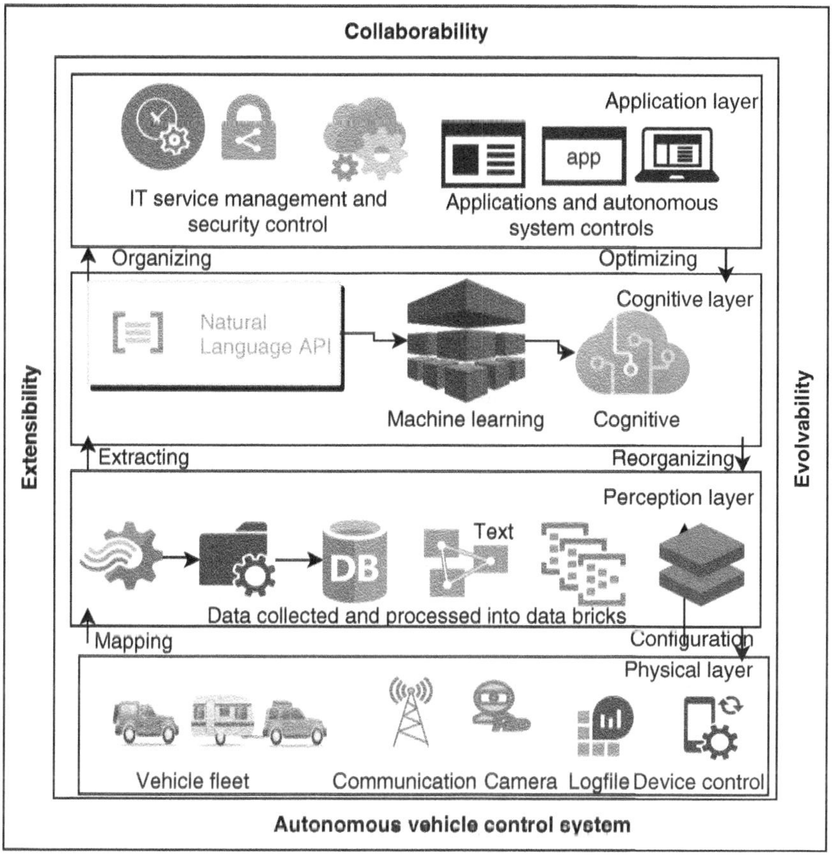


**Figure 6.8** Autonomous vehicle control system.

in an environment records live scenes to detect a position and perform analytics using sensor fusion technology. The middle layer perception process object, path, motion, and environment detection to record object list. The predictive model conducts behavior analysis.

Four-layer architecture in Figure 6.8 contains a physical layer, perception layer, cognitive layer, and application layer making up the proposed architecture. In order to maintain and control the devices, the physical layer contains device log details and a list of the scene's detected objects. After the data has been processed, the perception layer produces data bricks. This layer contains algorithms that clean the data and turn it into data bricks. The intelligence layer's primary function is to extract and organize information. The application layer utilizes this coordinated information to oversee different administrations and applications. At the top layer, security plays an important role.

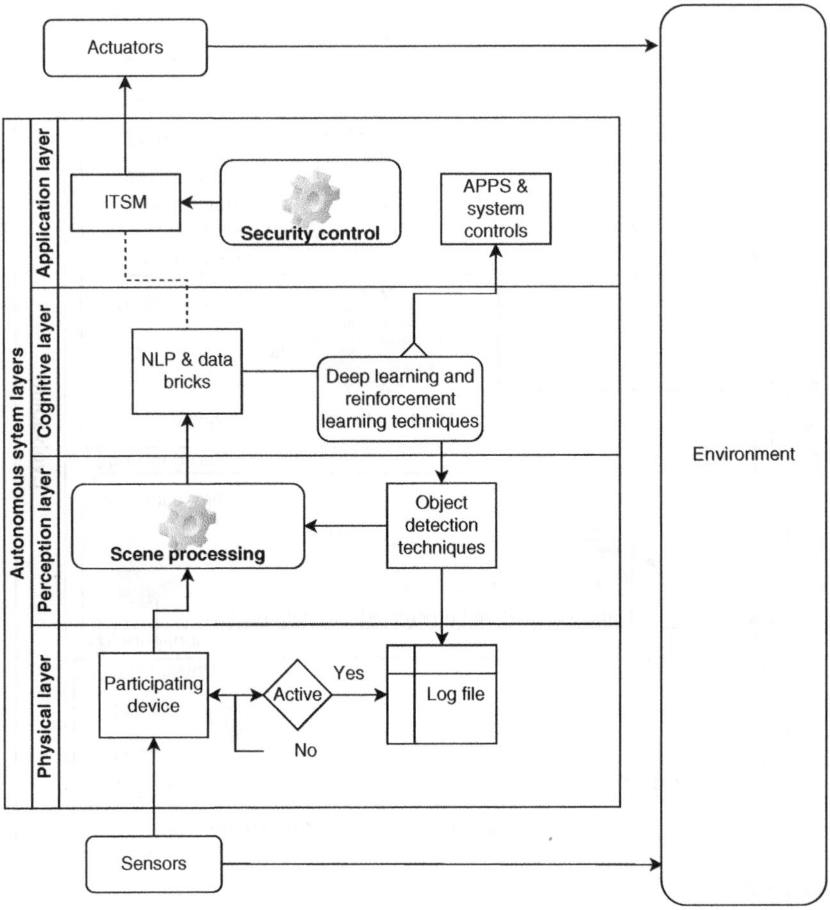


**Figure 6.9** Swimlane diagram for autonomous system.

Swimlane diagram in Figure 6.9 depicts the working model and helps to better understand the flow and to resolve integration challenges in autonomous systems.

### 6.4.1 Environmental Uncertainties

Error in perceived data or wrong input leads to failure of the system, and efficient error detection and correction leads to a better system. This is possible with the deployment of AI-based techniques used in autonomous systems for object detection and decision making are deep learning and reinforcement learning. It is a branch of machine learning that uses multilayered ANN to process huge amounts of data, while reinforcement learning is an area of machine learning concerned with the behavior of software agents acting in an environment to maximize rewards. System architecture assesses the potential use of state-of-art algorithms implemented on real-world problems. Environmental uncertainty like path, motion, and security challenges need to be overcome with the help of ITS and positioning systems. Designing an accurate positioning system is a must to handle various uncertainties during environment perception like pedestrian detection, path planning, motion control, and device security. To win user trust and to demonstrate great precision in their approach to tackling these challenges, incorporation of technologies must be accurate in functionality. It concludes that precise positioning solutions must be developed to handle a variety of uncertainties which is a key concern.

### 6.4.2 Security Challenges and Counter Measures

Data perceived from the environment should be reliable, clean, and safe to process. This is achieved with cleansing and preprocessing before implementation. Data is used to perform object detection with the support of machine learning models. These AI models are usually prone to security breaches and malicious attacks that impact the autonomous system performance. This raises the alarm and needs to study various sensor security issues and attacks that lead to system failures. Due to minimal synchronization between academic and industry implementation with respect to security against novel attacks is always an open problem. A survey with information about security attacks and their corresponding countermeasures, and autonomous vehicles targeted attack models [85] are discussed and concludes that efforts from CAV manufacturers to secure have no evidence to show that the current market-ready CAVs are ready to defend against current novel attacks. This acts as a basis for the research community to understand the current trend to improve. The use of blockchasin technology allows the exchange of information between connected devices rather than centralized network communication, reducing vulnerability.

## 6.5 Conclusion

Architecture is a major concern to handle any complexity in autonomous systems since they deal with multiple components interacting with the actual world while adhering to quality requirements, taking trade-offs into account, and making choices. This review is used to explain and list all possible challenges faced by autonomous systems. This evaluation can help experts and academics to recognize software and system design as crucial technological components. This is a hard problem to solve with diversified rules and behaviors in different parts of the world. A framework for an autonomous system with a four-layered architecture is provided with the intention to meet three major properties extensibility, evolvability, and cooperation. Layered architecture is a response to the challenges, and the number of layers in the suggested architecture in Figure 6.8 is a design to communicate with one another using communication technologies (B5G/5G/Wi-Fi/LTE), observe the surroundings using sensors, and gain complete autonomy using machine learning methods. This chapter aims to educate intended researchers about an autonomous system and its layered architecture, factors to consider while designing a secure and useful framework. As a consequence, reduced labor expenses and accurate and energy-efficient systems are into the ecosystem.

## References


**1** Kaiwartya, O., Abdullah, A.H., Cao, Y. et al. (2016). Internet of vehicles: motivation, layered architecture, network model, challenges, and future aspects. *IEEE Access* 4: 5356–5373. https://doi.org/10.1109/ACCESS.2016.2603219.

**2** Petit, J., Schaub, F., Feiri, M., and Kargl, F. (2015). Pseudonym schemes in vehicular networks: a survey. *IEEE Communications Surveys & Tutorials* 17 (1): 228–255. https://doi.org/10.1109/COMST.2014.2345420.

**3** Qazi, S., Sabir, F., Khawaja, B.A. et al. (2020). Why is internet of autonomous vehicles not as plug and play as we think ? Lessons to be learnt from present internet and future directions. *IEEE Access* 8: 133015–133033. https://doi.org/10.1109/ACCESS.2020.3009336.

**4** Arena, F. and Pau, G. (2019). An overview of vehicular communications. *Future Internet* 11 (2): 27. https://doi.org/10.3390/fi11020027.

**5** d'Orey, P.M. and Ferreira, M. (2014). ITS for sustainable mobility: a survey on applications and impact assessment tools. *IEEE Transactions on Intelligent Transportation Systems* 15 (2): 477–493. https://doi.org/10.1109/TITS.2013.2287257.

**6** Ondruš, J., Kolla, E., Vertaľ, P., and Šarić, Ž. (2020). How do autonomous cars work? *Transportation Research Procedia* 44: 226–233. https://doi.org/10.1016/j.trpro.2020.02.049.

**7** Ross, P. (2014). Robot, you can drive my car. *IEEE Spectrum* 51 (6): 60–90. https://doi.org/10.1109/MSPEC.2014.6821623.

**8** Alnema, Y.H., Almaged, M., and Noaman, M.N. (2020). Driverless model cars: a review and analysis of autonomous vehicle literature on technology and application. *International Review of Automatic Control (IREACO)* 13 (2): 84. https://doi.org/10.15866/ireaco.v13i2.18659.

9 Aria, M. and Cuccurullo, C. (2017). bibliometrix: An R-tool for comprehensive science mapping analysis. *Journal of Informetrics* 11 (4): 959–975. https://doi.org/10.1016/j.joi.2017.08.007.

10 Qiu, T., Zhao, Z., Zhang, T. et al. (2020). Underwater Internet of Things in smart ocean: system architecture and open issues. *IEEE Transactions on Industrial Informatics* 16 (7): 4297–4307. https://doi.org/10.1109/TII.2019.2946618.

11 (a) Zacharias, G. *Autonomous Horizons: The Way Forward.* Air University Press. (b) LeMay, C.E. (2019). *Center for Doctrine Development and Education.* Alabama: Maxwell Air Force Base.

12 Sethi, P. and Sarangi, S.R. (2017). Internet of Things: architectures, protocols, and applications. *Journal of Electrical and Computer Engineering* 2017: 9324035. https://doi.org/10.1155/2017/9324035.

13 Velasco-Hernandez, G., Yeong, D.J., Barry, J., and Walsh, J. (2020). Autonomous driving architectures, perception and data fusion: a review. *2020 IEEE 16th International Conference on Intelligent Computer Communication and Processing (ICCP)*, 315–321. Cluj-Napoca, Romania: IEEE. https://doi.org/10.1109/ICCP51029.2020.9266268. https://ieeexplore.ieee.org/document/9266268 (accessed 11 March 2023).

14 Campbell, S., O'Mahony, N., Krpalcova, L. et al. (2018). Sensor technology in autonomous vehicles: a review. *2018 29th Irish Signals and Systems Conference (ISSC)*, 1–4. Belfast: IEEE. https://doi.org/10.1109/ISSC.2018.8585340. https://ieeexplore.ieee.org/document/8585340 (accessed 11 March 2023).

15 Wang, Z., Wu, Y., and Niu, Q. (2020). Multi-sensor fusion in automated driving: a survey. *IEEE Access* 8: 2847–2868. https://doi.org/10.1109/ACCESS.2019.2962554.

16 Yeong, D.J., Velasco-Hernandez, G., Barry, J., and Walsh, J. (2021). Sensor and sensor fusion technology in autonomous vehicles: a review. *Sensors* 21 (6): 2140. https://doi.org/10.3390/s21062140.

17 Kumar, G.A., Lee, J.H., Hwang, J. et al. (2020). LiDAR and camera fusion approach for object distance estimation in self-driving vehicles. *Symmetry* 12 (2): 324. https://doi.org/10.3390/sym12020324.

18 Chen, H., Wen, Y., Zhu, M. et al. (2021). From automation system to autonomous system: an architecture perspective. *Journal of Marine Science and Engineering* 9 (6): 645. https://doi.org/10.3390/jmse9060645.

19 Elallid, B.B., Benamar, N., Hafid, A.S. et al. (2022). A comprehensive survey on the application of deep and reinforcement learning approaches in autonomous driving. *Journal of King Saud University, Computer and Information Sciences* 34 (9): 7366–7390. https://doi.org/10.1016/j.jksuci.2022.03.013.

20 Yurtsever, E., Lambert, J., Carballo, A., and Takeda, K. (2020). A survey of autonomous driving: common practices and emerging technologies. *IEEE Access* 8: 58443–58469. https://doi.org/10.1109/ACCESS.2020.2983149.

**21** Parekh, D., Poddar, N., Rajpurkar, A. et al. (2022). A review on autonomous vehicles: progress, methods and challenges. *Electronics* 11 (14): 2162. https://doi.org/10.3390/electronics11142162.

**22** Chhabra, R., Prabaharan, B., and Aggarwal, R. Role of wireless body area networks in intelligent transportation systems: a survey. *International Journal of Sensors, Wireless Communications and Control* 11 (8): 800–813.

**23** Ahangar, M.N., Ahmed, Q.Z., Khan, F.A., and Hafeez, M. (2021). A survey of autonomous vehicles: enabling communication technologies and challenges. *Sensors* 21 (3): 706. https://doi.org/10.3390/s21030706.

**24** Giacalone, J.-P., Bourgeois, L., and Ancora, A. (2019). Challenges in aggregation of heterogeneous sensors for Autonomous Driving Systems. *2019 IEEE Sensors Applications Symposium (SAS)*, 1–5. Sophia Antipolis, France: IEEE. https://doi.org/10.1109/SAS.2019.8706005. https://ieeexplore.ieee.org/document/8706005 (accessed 11 March 2023).

**25** Wang, J., Wang, C., Lin, Q. et al. (2022). Adversarial attacks and defenses in deep learning for image recognition: a survey. *Neurocomputing* 514: 162–181. https://doi.org/10.1016/j.neucom.2022.09.004.

**26** Fayyad, J., Jaradat, M.A., Gruyer, D., and Najjaran, H. (2020). Deep learning sensor fusion for autonomous vehicle perception and localization: a review. *Sensors* 20 (15): 4220. https://doi.org/10.3390/s20154220.

**27** Dai, X., Yuan, X., Pei, L., and Wei, X. (2020). Deeply supervised Z-style residual network devotes to real-time environment perception for autonomous driving. *IEEE Transactions on Intelligent Transportation Systems* 21 (6): 2396–2408. https://doi.org/10.1109/TITS.2019.2918227.

**28** Ma, X., Niu, Y., Gu, L. et al. (2021). Understanding adversarial attacks on deep learning based medical image analysis systems. *Pattern Recognition* 110: 107332. https://doi.org/10.1016/j.patcog.2020.107332.

**29** Kuutti, S., Bowden, R., Jin, Y. et al. (2021). A survey of deep learning applications to autonomous vehicle control. *IEEE Transactions on Intelligent Transportation Systems* 22 (2): 712–733. https://doi.org/10.1109/TITS.2019.2962338.

**30** Pham, M. and Xiong, K. (2021). A survey on security attacks and defense techniques for connected and autonomous vehicles. *Computers & Security* 109: 102269. https://doi.org/10.1016/j.cose.2021.102269.

**31** Mekrache, A., Bradai, A., Moulay, E., and Dawaliby, S. (2022). Deep reinforcement learning techniques for vehicular networks: recent advances and future trends towards 6G. *Vehicular Communications* 33: 100398. https://doi.org/10.1016/j.vehcom.2021.100398.

**32** Garg, A. and Sood, R. (2022). Application centric mobility framework for next generation networks. *Wireless Personal Communications* 125 (2): 1243–1259. https://doi.org/10.1007/s11277-022-09599-w.

**33** Jain, V., Saini, D., Gupta, M. et al. (2021). A comprehensive review on design of autonomous robotic boat for rescue applications. *Mathematical Problems in Engineering* 2021: 6614002. https://doi.org/10.1155/2021/6614002.

**34** Magnussen, A.F., Le, N., Hu, L., and Wong, W.E. (2020). A survey of the inadequacies in traffic sign recognition systems for autonomous vehicles. *International Journal of Performability Engineering* 16 (10): 1588. https://doi.org/10.23940/ijpe.20.10.p10.15881597.

**35** Zhang, R., Li, S., Ji, G. et al. (2021). Survey on deep learning-based marine object detection. *Journal of Advanced Transportation* 2021: 5808206. https://doi.org/10.1155/2021/5808206.

**36** Badue, C., Guidolini, R., Carneiro, R.V. et al. (2021). Self-driving cars: a survey. *Expert Systems with Applications* 165: 113816. https://doi.org/10.1016/j.eswa.2020.113816.

**37** Manzoor, S., Joo, S.-H., Kim, E.-J. et al. (2021). 3D recognition based on sensor modalities for robotic systems: a survey. *Sensors* 21 (21): 7120. https://doi.org/10.3390/s21217120.

**38** Nacpil, E.J.C., Wang, Z., and Nakano, K. (2021). Application of physiological sensors for personalization in semi-autonomous driving: a review. *IEEE Sensors Journal* 21 (18): 19662–19674. https://doi.org/10.1109/JSEN.2021.3100038.

**39** Zhang, Y., Carballo, A., Yang, H., and Takeda, K. (2023). Perception and sensing for autonomous vehicles under adverse weather conditions: a survey. *ISPRS Journal of Photogrammetry and Remote Sensing* 196: 146–177. https://doi.org/10.1016/j.isprsjprs.2022.12.021.

**40** Lopez, A., Malawade, A.V., Al Faruque, M.A. et al. (2019). Security of emergent automotive systems: a tutorial introduction and perspectives on practice. *IEEE Design & Test* 36 (6): 10–38. https://doi.org/10.1109/MDAT.2019.2944086.

**41** De Sousa, L.S. and Drummond, A.C. (2023). Metropolitan optical networks: a survey on single-layer architectures. *Optical Switching and Networking* 47: 100719. https://doi.org/10.1016/j.osn.2022.100719.

**42** Camara, F., Bellotto, N., Cosar, S. et al. (2021). Pedestrian models for autonomous driving Part I: low-level models, from sensing to tracking. *IEEE Transactions on Intelligent Transportation Systems* 22 (10): 6131–6151. https://doi.org/10.1109/TITS.2020.3006768.

**43** Li, P., Nguyen, A.-T., Du, H. et al. (2021). Polytopic LPV approaches for intelligent automotive systems: state of the art and future challenges. *Mechanical Systems and Signal Processing* 161: 107931. https://doi.org/10.1016/j.ymssp.2021.107931.

**44** Laconte, J., Kasmi, A., Aufrère, R. et al. (2021). A survey of localization methods for autonomous vehicles in highway scenarios. *Sensors* 22 (1): 247. https://doi.org/10.3390/s22010247.

**45** Gomes, T., Matias, D., Campos, A. et al. (2023). A survey on ground segmentation methods for automotive LiDAR sensors. *Sensors* 23 (2): 601. https://doi.org/10.3390/s23020601.

**46** Zhang, J. and Letaief, K.B. (2020). Mobile edge intelligence and computing for the internet of vehicles. *Proceedings of the IEEE* 108 (2): 246–261. https://doi.org/10.1109/JPROC.2019.2947490.

**47** Alaba, S.Y. and Ball, J.E. (2022). A survey on deep-learning-based LiDAR 3D object detection for autonomous driving. *Sensors* 22 (24): 9577. https://doi.org/10.3390/s22249577.

**48** Khoshavi, N., Tristani, G., and Sargolzaei, A. (2021). Blockchain applications to improve operation and security of transportation systems: a survey. *Electronics* 10 (5): 629. https://doi.org/10.3390/electronics10050629.

**49** Mohammed, A.S., Amamou, A., Ayevide, F.K. et al. (2020). The perception system of intelligent ground vehicles in all weather conditions: a systematic literature review. *Sensors* 20 (22): 6532. https://doi.org/10.3390/s20226532.

**50** Mohamed, A., Ren, J., Gindy, M.E. et al. (2018). Literature survey for autonomous vehicles: sensor fusion, computer vision, system identification and fault tolerance. *International Journal of Automation and Control* 12 (4): 555. https://doi.org/10.1504/IJAAC.2018.095104.

**51** Wei, S., Yu, D., Lin Guo, C. et al. (2019). Survey of connected automated vehicle perception mode: from autonomy to interaction. *IET Intelligent Transport Systems* 13 (3): 495–505. https://doi.org/10.1049/iet-its.2018.5239.

**52** Mostaani, A., Vu, T.X., Sharma, S.K. et al. (2022). Task-oriented communication design in cyber-physical systems: a survey on theory and applications. *IEEE Access* 10: 133842–133868. https://doi.org/10.1109/ACCESS.2022.3231039.

**53** Sanchez, P.J.B., Márquez, F.P.G., Govindara, S. et al. (2021). Use of UIoT for offshore surveys through autonomous vehicles. *Polish Maritime Research* 28 (3): 175–189. https://doi.org/10.2478/pomr-2021-0044.

**54** Alghodhaifi, H. and Lakshmanan, S. (2021). Autonomous vehicle evaluation: a comprehensive survey on modeling and simulation approaches. *IEEE Access* 9: 151531–151566. https://doi.org/10.1109/ACCESS.2021.3125620.

**55** Shi, W., Alawieh, M.B., Li, X., and Yu, H. (2017). Algorithm and hardware implementation for visual perception system in autonomous vehicle: a survey. *Integration* 59: 148–156. https://doi.org/10.1016/j.vlsi.2017.07.007.

**56** Belwafi, K., Alkadi, R., Alameri, S.A. et al. (2022). Unmanned aerial vehicles' remote identification: a tutorial and survey. *IEEE Access* 10: 87577–87601. https://doi.org/10.1109/ACCESS.2022.3199909.

**57** Devi, S., Malarvezhi, P., Dayana, R., and Vadivukkarasi, K. (2020). A comprehensive survey on autonomous driving cars: a perspective view. *Wireless Personal Communications* 114 (3): 2121–2133. https://doi.org/10.1007/s11277-020-07468-y.

**58** Banafaa, M., Shayea, I., Din, J. et al. (2023). 6G Mobile communication technology: requirements, targets, applications, challenges, advantages, and opportunities. *Alexandria Engineering Journal* 64: 245–274. https://doi.org/10.1016/j.aej.2022.08.017.

**59** Chowdhury, A., Karmakar, G., Kamruzzaman, J. et al. (2020). Attacks on self-driving cars and their countermeasures: a survey. *IEEE Access* 8: 207308–207342. https://doi.org/10.1109/ACCESS.2020.3037705.

**60** Dai, D., Chen, Z., Bao, P., and Wang, J. (2021). A review of 3D object detection for autonomous driving of electric vehicles. *World Electric Vehicle Journal* 12 (3): 139. https://doi.org/10.3390/wevj12030139.

**61** Ridao, P., Carreras, M., Ribas, D. et al. (2014). Intervention AUVs: the next challenge. *IFAC Proceedings Volumes* 47 (3): 12146–12159. https://doi.org/10.3182/20140824-6-ZA-1003.02819.

**62** Wang, Z., Wei, H., Wang, J. et al. (2022). Security issues and solutions for connected and autonomous vehicles in a sustainable city: a survey. *Sustainability* 14 (19): 12409. https://doi.org/10.3390/su141912409.

**63** Kloukiniotis, A., Papandreou, A., Lalos, A. et al. (2022). Countering adversarial attacks on autonomous vehicles using denoising techniques: a review. *IEEE Open Journal of Intelligent Transportation Systems* 3: 61–80. https://doi.org/10.1109/OJITS.2022.3142612.

**64** Balkus, S.V., Wang, H., Cornet, B.D. et al. (2022). A survey of collaborative machine learning using 5G vehicular communications. *IEEE Communications Surveys & Tutorials* 24 (2): 1280–1303. https://doi.org/10.1109/COMST.2022.3149714.

**65** Rodríguez, E., Otero, B., and Canal, R. (2023). A survey of machine and deep learning methods for privacy protection in the Internet of Things. *Sensors* 23 (3): 1252. https://doi.org/10.3390/s23031252.

**66** Schwarting, W., Alonso-Mora, J., and Rus, D. (2018). Planning and decision-making for autonomous vehicles. *Annual Review of Control, Robotics, and Autonomous Systems* 1 (1): 187–210. https://doi.org/10.1146/annurev-control-060117-105157.

**67** Romero, L.M., Guerrero, J.A., and Romero, G. (2021). Road curb detection: a historical survey. *Sensors* 21 (21): 6952. https://doi.org/10.3390/s21216952.

**68** Riedmaier, S., Ponn, T., Ludwig, D. et al. (2020). Survey on scenario-based safety assessment of automated vehicles. *IEEE Access* 8: 87456–87477. https://doi.org/10.1109/ACCESS.2020.2993730.

**69** Zhou, T., Yang, M., Jiang, K. et al. (2020). MMW radar-based technologies in autonomous driving: a review. *Sensors* 20 (24): 7283. https://doi.org/10.3390/s20247283.

**70** Malik, S., Khan, M.A., El-Sayed, H. et al. (2022). How do autonomous vehicles decide? *Sensors* 23 (1): 317. https://doi.org/10.3390/s23010317.

**71** Sulieman, N.A., Ricciardi Celsi, L., Li, W. et al. (2022). Edge-oriented computing: a survey on research and use cases. *Energies* 15 (2): 452. https://doi.org/10.3390/en15020452.

**72** Baruah, S., Lee, P., Sarathy, P., and Wolf, M. (2020). Achieving resiliency and behavior assurance in autonomous navigation: an industry perspective. *Proceedings of the IEEE* 108 (7): 1196–1207. https://doi.org/10.1109/JPROC.2020.2978661.

**73** Olakanmi, O.O. (2020). SAPMS: A secure and anonymous parking management system for autonomous vehicles. *International Journal of Information and Computer Security* 12 (1): 20–39. https://doi.org/10.1504/IJICS.2020.103996.

**74** Chahal, R.K., Kumar, N., and Batra, S. (2020). Trust management in social Internet of Things: a taxonomy, open issues, and challenges. *Computer Communications* 150: 13–46. https://doi.org/10.1016/j.comcom.2019.10.034.

**75** Liu, S., Liu, H., John, V. et al. (2020). Enhanced situation awareness through CNN-based deep multimodal image fusion. *Optical Engineering* 59 (05): 053103. https://doi.org/10.1117/1.OE.59.5.053103.

**76** Ansari, F., Hold, P., and Khobreh, M. (2020). A knowledge-based approach for representing jobholder profile toward optimal human–machine collaboration in cyber physical production systems. *CIRP Journal of Manufacturing Science and Technology* 28: 87–106. https://doi.org/10.1016/j.cirpj.2019.11.005.

**77** Miglani, A. and Kumar, N. (2019). Deep learning models for traffic flow prediction in autonomous vehicles: a review, solutions, and challenges. *Vehicular Communications* 20: 100184. https://doi.org/10.1016/j.vehcom.2019.100184.

**78** Dai, Y. and Lee, S.-G. (2020). Perception, planning and control for self-driving system based on on-board sensors. *Advances in Mechanical Engineering* 12 (9): 168781402095649. https://doi.org/10.1177/1687814020956494.

**79** Jain, V., Patnaik, S., Vlădicescu, F.P., and Sethi, I.K. (ed.) (2020). Recent trends in intelligent computing, communication and devices: proceedings of ICCD 2018. In: *Advances in Intelligent Systems and Computing*, vol. 1006. Singapore: Springer https://doi.org/10.1007/978-981-13-9406-5. http://link.springer.com/10.1007/978-981-13-9406-5.

**80** Qian, R., Lai, X., and Li, X. (2022). 3D object detection for autonomous driving: a survey. *Pattern Recognition* 130: 108796. https://doi.org/10.1016/j.patcog.2022.108796.

**81** Das, S., Mandal, S., Bhoyar, A. et al. (2020). Multi-criteria online frame-subset selection for autonomous vehicle videos. *Pattern Recognition Letters* 133: 349–355. https://doi.org/10.1016/j.patrec.2020.03.031.

**82** Liu, D., Cui, Y., Chen, Y. et al. (2020). Video object detection for autonomous driving: motion-aid feature calibration. *Neurocomputing* 409: 1–11. https://doi.org/10.1016/j.neucom.2020.05.027.

**83** Krishnamurthi, R., Kumar, A., and Gill, S.S. (2022). *Autonomous and Connected Heavy Vehicle Technology*. Intelligent Data Centric Systems. London, San Diego, CA, Cambridge, MA, Kidlington, Oxford: Academic Press, an imprint of Elsevier.

**84** Siddiqui, M.U.A., Qamar, F., Tayyab, M. et al. (2022). Mobility management issues and solutions in 5G-and-beyond networks: a comprehensive review. *Electronics* 11 (9): 1366. https://doi.org/10.3390/electronics11091366.

**85** Kumar, A., Sharma, S., Singh, A. et al. (2021). Revolutionary strategies analysis and proposed system for future infrastructure in Internet of Things. *Sustainability* 14 (1): 71. https://doi.org/10.3390/su14010071.

# 7

# Integrated Trends, Opportunities, and Challenges of 5G and Internet of Things

*Ekta Dixit*[1] *and Shalli Rani*[2]

[1] *Department of Computer Science, S.S.D Women's Institute of Technology, Bathinda, Punjab, India*
[2] *Chitkara University Institute of Engineering and Technology, Chitkara University, Rajpura, Punjab, India*

## 7.1 Introduction

The world has altered and evolved over the past 10 years due to the introduction of several new technologies, such as IoT and big data, as well as 5G cellular networks [1]. Regarding Internet of Things deployments, IOT is a worldwide network based on superior communication protocols that utilises diverse technologies to collect and provide physical world observation data. IOT modules will have the capacity to connect over 80 billion users by 2030. It produces vast amounts of data that are more heterogeneous, faster, have more nodes, are larger in size, and have more features. In the meantime, 5G networks are now recognised as a crucial IOT development driver. The 5G network is expected to provide billions of sensors with Internet connections by delivering broad coverage, higher output, reduced latency, and massive network link capacity. Actually, a few promising technologies, such as Machine-to-Machine (M2M) communications, could increase the adaptability of 5G networks [2]. Thus, uniform and heterogeneous sensor networks are able to link large numbers of devices that can sense and play a major role in providing people with sophisticated and smart services. People are being encouraged to specifically comprehend the opportunities, risks, and conceptual framework of these technologies. IOT technology has a direct impact on computational complexity and costs in terms of data processing and storage, and it raises new issues for 5G and IOT, such as safety, security, and privacy given their distinct features. Simultaneously, the platform's ability to communicate and be used are both challenging in different situations. Intelligent techniques

*Current and Future Cellular Systems: Technologies, Applications, and Challenges*, First Edition.
Edited by Garima Chopra, Suhaib Ahmed, and Shalli Rani.

like machine learning and insightful learning are therefore necessary to handle large amounts of data and improve the communication channel [3]. The IOT 5G model is merging of technology, the cloud, and other things. An example of an IOT model is one that includes several Internet-based network connection actors and other elements. When intelligence and other components are merged, high-functioning agents or tools are produced that can handle challenging tasks including cognitive control, natural language processing, and recognition of objects. The intellect is a person of intelligence. Actually, promising strategies for reaching these goals include profound learning and machine learning. Based on a few essential technologies like 5G, communication protocols and data interchange quality such as the convergence of internet capabilities and information, the intelligent internet offers improved networking capabilities In conclusion, the author concentrated on the 5G idea in relation to IOT, which includes the internet, intelligence, and various devices such as processors, processing centers, and sensing regions. Artificial intelligence (AI) techniques to process 5G with IOT are meant to be made possible by the processing center [4]. Large volumes of data are continuously processed electronically by the processing center. Sensing regions and processing center are connected by the object processor, which also gathers and sends raw data from the sensing areas to the processing centers. In order to function effectively processing center send information to the object processor. Large sensors are used in the sensing zones to gather unprocessed data.

This chapter clarifies the core idea, design, features, and development of the 5G ecosystem, an outline of the Internet of Things and its features, IOT three layered architecture, A list of prerequisites is provided for creating an integrated 5G IoT ecosystem. The 5G integrated IoT ecosystem's prospects and difficulties discussed.

### 7.1.1 Overview of 5G

The upcoming generations of wireless networks can advance wireless communication exponentially due to 5G. The needs of several applications are growing daily. Wireless technology is evolving quickly, and it is always attempting to stay updated. Applications that rely on big data and other networks for various purposes, such as expedited money transfers, early identification of serious illnesses, inventory management, etc., make advantage of 5G's features to improve the system and boost effectiveness [5]. The newest apps, which demand high data processing capacity with high quality of service (QoS) and quality of experience (QoE), cannot be supported by 4G and earlier generations. Similarly, the transmission rates are expressed in Gbps, with a typical rate of 100+ Mbps. According to tests, 5G can transmit data at up to 20 Gbps, which is 100 times quicker than 4G. Thus, implementing 5G in the network infrastructure has the potential to completely alter how people communicate.

### 7.1.2 Evolution from 1G to 5G

The introduction of the first portable (1G) was between 1970 and 1980. Digital technology, which made use of SMS efficiency following the development of the microprocessor, fueled the second generation of radio communication (2G). Once a few years have passed, phone conversations, MMS, image sharing, and other functions may be shared more easily due to the combination of GPRS and 2G network. In the twenty-first century, third-generation (3G) sound and notification devices, such as live TV and fast internet, were introduced for better systems. The extraordinarily high velocity of internet connections in 4G-capable devices prompted the development of the Fourth Generation (4G). The barriers of mobile phones are no longer sufficient for current communication, which also involves laptops, iceboxes, cars, and other contemporary devices in addition to architecture. In order to comprehend continuous device interaction and access other devices, a quick site and data transfer feature is required. This fact also stimulates the fifth generation (5G) of communication innovation [6]. The evolution of technology are presented in Figure 7.1.

### 7.1.3 5G Architecture

5G's network architecture uses a sophisticated dispersed design and radio access network (RAN) points that are virtual, flexible, and adaptive in place of base stations (BS) [7]. The 5G architecture have following functions.

**RAN:** 5G connects many technologies that provide FDD frequency through the usage of Radio Access Networks (RAN).

**Data Network:** This service offers internet connection through third-party operators.

**Access and Mobility Management:** This feature protects the integrity, grants access, controls mobility, establishes connections between devices, and more.

**Control Policy:** This feature starts the foundations for policies that regulate network activity.

**Network Slice Selection:** This feature selects user equipment instances and data for the help function.

Many more features of the 5G architecture exist, and they differ depending on the application. Nonetheless, the fundamental goal of every function is to attain increased performance, minimal latency, appropriate handling, and security [8].

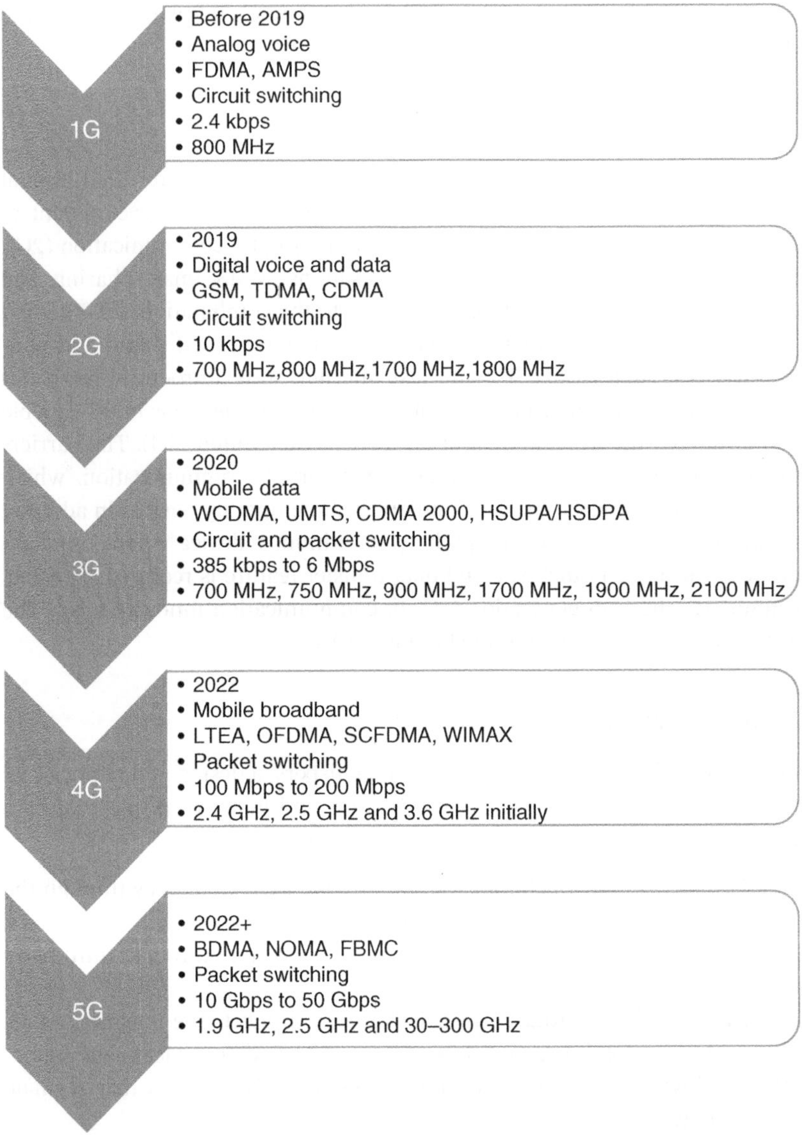


**Figure 7.1** Evolution of technology.

### 7.1.4 Overview of IoT

The A networked digital system made up of various electrical gadgets such as data-computing nodes, sensors, actuators, receivers is called the Internet of Things, or IoT. IoT devices have revolutionised the data collection and processing system by removing the need for human intervention. Smart home, smart car, smart agriculture, and many more systems are all improved by IoT devices. They have been employed to carry out, keep an eye on, and generate responses in response to the data acquired. On the other side, the Internet of Things expands and improves network technology founded on current internet technology, enabling communication and connection between computers and smart items [9]. Any object that uses the Internet to generate, exchange, and communicate data with other things for the purposes of intelligent recognition, orientation tracing, tracking, and management is commonly referred to as the Internet of Things. Numerous sensors and peripherals, including RFID, heat sensors, GPS, and others, carry out this procedure [10].

### 7.1.5 Features of IoT

In order to build the infrastructure there are many IOT requirements. Here, we'll talk about some of the most advantageous aspects of IoT.

#### 7.1.5.1 Avalability

IoT availability needs to be addressed at the hardware and software levels in order to give clients access to facilities whenever and wherever they require them. Software availability refers to an IoT system's ability to provide functionality to every user, anywhere [11]. Hardware availability describes a computer's capacity to constantly work with Internet of Things capabilities and protocols. Protocols must be deployed inside the single board resource's restricted devices in order to enable IoT capabilities. Ensuring the availability of vital equipment and services is one method to achieve high IoT service availability [12].

#### 7.1.5.2 Mobility

Accessibility issues prevent most utilities from being implemented in the Internet of Things, even if they are intended to be delivered through smartphones. Preserving customers' favourite resources even when they move is a fundamental tenet of the Internet of Things. Service disruptions may occur if mobile devices move from one gateway to another.

#### 7.1.5.3 Scalabilty

Within the Internet of Things, scalability is the capacity to integrate additional client hardware, software, and functionalities without sacrificing the effectiveness of current systems. Adding more processes and managing additional devices is not easy, particularly when various hardware platforms and communication protocols need to be taken into account.

#### 7.1.5.4 Security

The data exchange between billions devices, offers the foundation for the basic operation of the Internet of Things. Sharing of keys among devices, not considered by standards is a significant issue with IoT security. Transparent and straightforward access control management—such as allowing one vendor to view the data—is required due to the increasing number of smart devices in our environment that contain sensitive data. However, another person operates the apparatus [13].

#### 7.1.5.5 Context Awareness

Context awareness pertains to the capacity of Internet of Things (IoT) systems and devices to comprehend and react to the context and surroundings in which they are functioning. Sensors and other technologies that can detect and gather information about the surroundings are used to do this. An essential component of the Internet of Things is context awareness, which allows devices and systems to make decisions and take action based on the environment in which they are used.

### 7.1.6 IoT Architecture

Different IoT world concepts have been proposed. These structures can be broadly categorised into three groups. Three-layer, four-layer, and five-layer architectures are the three different forms of architecture. Three layer architecture results in following layers.

#### 7.1.6.1 Application Layer

This layer may contain a variety of services depending on the implementation. Examples of IoT deployment in smart homes and cities comprise smart grids, and healthcare. Customers can get the services they require from the application layer. For example, the application layer should provide the client requesting the information with the humidity and temperature values. Because it makes it possible to develop superior smart services that satisfy user needs, this layer is essential to the Internet of Things.

#### 7.1.6.2 Network Layer

The network layer regulates data transit to lower layers by serving as a bridge. With the visual layer, this layer is connected. The various protocols and standards are followed when connecting various smart devices to the network layer.

This layer is made possible by smart gadgets, the Internet protocol, and server backend architecture. Furthermore, the network layer can be managed based on the specific requirements of the environment that is being implemented. Data communication is extremely vulnerable to cyberattacks. The most widely used, along with the most recent adoption of blockchain technology, is an intelligent intrusion detection key encryption with protected management-based Internet of Things security framework.

#### 7.1.6.3 Edge Layer

In order to communicate with clients or their working domains, the edge layer uses gateways in a coordinating role to deal with IoT devices or sensors such as RFID, various actuators, cameras, etc. Gathering data from the environment and sending it on for additional processing is its primary responsibility. in order to comply with protocol translation and traffic control IPs such as IPv6 and gateways are used. A typical security method cannot safeguard these devices due to the sensors and actuators. Therefore, a few hazards to security are made visible by device compatibility and physical accessibility.

## 7.2 Requirements for Integration of 5G with IoT

5G capable IoT requires particular consideration because to its diversity, development, and use. Nonetheless, any architecture must adhere to a few standards [1].

In light of the delicate nature of the internet and the implications for health, 5G IoT must guarantee a low latency of 1 ms.

Low energy consumption for IoT devices with short battery lives and sufficient energy for 5G transfer of data must be guaranteed by the architecture.

Future needs for speed—a high of 25 Mbps—must be accommodated.

With so much data being transmitted at such a rapid rate, security needs to be excellent.

### 7.2.1 Integrated 5G IoT Layered Architecture

IoT is connected to data gathering, processing, analysis, and sharing between devices and communication networks. It is comprised of the primary four-layer architecture within the 5G framework [14].

i) **Thing Layer:** This layer interacts with the network layer and consists of physical systems including actuators, gadgets, and sensors.
ii) **The Network Layer The Network Layer is Divided into two Sublayers:** (1) low power wide area networks (LPWANs), like SigFox, LoRa, ZigBee, and NB-IoT; and (2) 5G backhaul-based connectivity.

iii) **Middleware Layer:** the middle layer is regarded as the network's brains. The deployment of fog computing, edge computing, cloud computing, AI vision, and big data analytics highlights the IoT framework's emphasis on cutting-edge technology and solutions [15].
iv) **Application Layer:** this layer displays IoT applications that are implemented in many fields, such as traffic systems, agriculture, industries and buildings, and IoT ecosystems. This layer combines all applications, technologies, and techniques needed to communicate with people online.

## 7.3 Opportunities of 5G integrated IoT

The rapidly developing technology referred to as 5G has created a plethora of new prospects. With enormous bandwidth and high speed, over 60,000 links will be supported. In addition, 5G unifies all networks onto a single platform. Additionally, it allows users to keep an eye on their accounts and respond quickly. 5G networks can be seamlessly integrated with previous network generations. Furthermore, 5G is intended to provide a continuous and uninterrupted connectivity everywhere. Connecting IoT to 5G will speed up development in numerous other areas, such as business, industry, technology, etc.

### 7.3.1 Smart Cities

5G will be more widely used in smart city initiatives, ranging from supply chain management of all requirements and everyday demands to home automation systems and enhanced communication. 5G networks with advanced sensors that improve city infrastructure will benefit smart cities. Massive data management and the integration of several intelligent technologies will be made possible by 5G, which will further strengthen the bonds that bind a truly connected city [16].

### 7.3.2 Smart Vehicles

Automated vehicles use sophisticated sensors to gather a vast amount of data on a variety of topics, including traffic, weather, temperature, GPS position, and other variables. When so much data is generated and processed, a lot of energy is used. These cars mostly rely on real-time data transmission to provide the finest services. With fast internet and low latency, the system integrated into these cars may be set up to gather all necessary data, even those that are vital. In due course, it will allow the cars to independently observe how it operates and improve subsequent models, such as the system algorithm [11].

### 7.3.3 Device to Device Communications

5G network facilitates seamless direct-to-direct (D2D) communication, eliminating the need for messengers [17]. There are four different ways that devices can communicate with each other using D2D technology: directly through an operator-controlled link, directly through a device-controlled link, transmission among devices etc [18].

### 7.3.4 Business

5G capable Beyond merely advancing technology, Internet of Things is expected to support 22 million jobs worldwide. This increase is probably going to be driven by the modernization of transportation, industry, agriculture, and other physical industries. 5G has the potential to enhance intelligent machinery and smart production. IoT will soon be able to control network traffic almost instantly, resolve issues, improve security and public safety, and function remotely due to 5G [19].

### 7.3.5 Satelite and Aerial Research

A wide range of opportunities for satellite and aerial communication and research are also made possible by the development of the 5G network. Research on High Altitude Performance Systems is being conducted in conjunction with satellite technology [20]. It becomes a smooth integration when Narrowband-IoT (NB-IoT) is added to the 5G network. It can facilitate low-bitrate communication between satellites with intermediate structural characteristics. Combining terrestrial and satellite networks can potentially improve the 5G mobile network [21].

### 7.3.6 Video Surveillance

Surveillance systems are used globally by many organisations and private sectors. Many modern video surveillance systems still employ wired connections. The popularity of wireless communications, such as Wi-Fi and cellular, continues to increase because of its low cost of implementation, ease of use, and speed. It is anticipated to rise faster in the near future in order to provide greater security [22].

## 7.4 Challenges of 5G Integrated IoT

The pervasive integration of both technologies given us access to remarkable features. Cutting-edge research has demonstrated that increased connectivity that guarantees dependable Quality of Service (QoS) can lead to more applications and developments. There are many additional issues that need to be resolved as a result of these developments. We'll talk about a few of the major obstacles to 5G integrated IoT here.

### 7.4.1 Insufficient Control over Data Storage and Usage

A 5G IoT network generates enormous amounts of data. It is produced using gadgets that are exclusive to certain industries. Because of this, these data are usually beyond the control of all parties involved, including users [23]. Through common 5G network, the data may be shared by all service providers, all suppliers whose equipment generates a network node, or all users using a single cloud platform. Tracking which data originate from whom, as well as the identity of the author and processing system, is therefore a challenging undertaking [24].

### 7.4.2 Scalability

Cloud-based architecture makes it possible for 5G IoT to oversee and operate the network as a whole. Stated differently, the network's nodes generate data that is processed and sent to a shared cloud. The network nodes then relay the control signals in order to perform various functions, including routing, traffic management, fault tolerance, and storage reallocation. Nevertheless, the amount of data generated by connected devices increases along with their quantity, which poses a growing challenge for centralised cloud servers in terms of capacity and processing power. A gateway or an edge node is also used by devices in the 5G and IoT ecosystems to connect to cloud nodes. Upon receiving a large number of requests for connections to the cloud, the fronthaul, midhaul, and backend networks instantly terminate [25].

### 7.4.3 Heterogeneity of 5G and IoT Data

The diverse nature of the both technologies raises a number of interoperability issues when developing different applications. It is nearly hard to predict precise features and results due to the diversity and multidimensionality of the data generated in both networks. Consequently, in order to train this kind of data, early steps like preprocessing, cleaning, and sorting are needed because integrating such a varied collection of data can result in inaccurate calculations. Therefore, greater consideration must be given to this factor while testing datasets in various scenarios in order to train datasets and choose features [26].

### 7.4.4 Blockchain Processing Time

In the blockchain ecosystem, a few self-executing tasks are needed to develop the chain, such as transaction and block verification. Ensuring the authenticity of blocks in a blockchain requires these computations to adhere to specified cryptographic protocols, which can be time-consuming. While it might be a way to reduce the quantity of training data, there are particular computing constraints in the Internet of Things setting that could cause security issues. Because of this, a major issue with implementing blockchain in the context of IoT is that it requires fewer replacements that are resource-intensive in order to shorten processing times.

### 7.4.5 5G mm-Wave Issues

Low latency, larger capacity allocation for Multiple Input Multiple Output (MIMO) technologies, and increased robustness are all potential benefits of 5G mm-wave use. However, there is a communication range constraint. Furthermore, 5G mm-wave devices are integrated with LTE and mm-wave band devices, which have distinct features. Once more, obstructions can limit its movement in its nutrient-rich scattering environment. Because it uses a lot of gear, it also uses a lot of power. Therefore, it is difficult to use IoT infrastructure when there is a disruption in the weather [27].

### 7.4.6 Threat Protection of 5G IoT

In a 5G integrated IoT architecture, security vulnerabilities will resurface even after specific 5G and IoT issues have been mitigated. Numerous security risks, such as denial-of-service (DoS) attacks, network failures, malware, penetration attempts, Man-in-the-middle (MITM) attacks, TCP level attacks, security key exposure resets, IP spoofing attacks, and so on, are targeted at technologies such as SDN, NFV, cloud servers, etc [28]. These attacks have the potential to significantly affect data privacy in addition to exposing the entire 5G IoT application. Certain threats can be prevented by using low-power nodes and sensors. However, it remains an issue for potential 5G IoT in the future [29].

## 7.5 Conclusion

The possibilities and difficulties of integrating 5G IoT in several application sectors are the key topics of this chapter, along with potential solutions. The 5G network is expected to revolutionise the Internet of Things by connecting over 100 billion devices and exchanging data at a rate at least 10 times faster than LTE. The goal should be to introduce a 5G IoT network architecture that is scalable, economical, and efficient, capable of connecting enormous numbers of devices. There are several potential and obstacles associated with using 5G IoT in various areas to boost efficacy. In this chapter, we attempt to examine how the 5G integrated IoT ecosystem can eliminate IoT's limitations while outlining its strengths. The design of the 5G Internet of Things also demonstrates the numerous obstacles and future directions for study. We anticipate that these examples will offer in-depth understanding and stimulate the creation of 5G IoT applications.

## References


**1** Li, S., Da Xu, L., and Zhao, S. (2018). 5G internet of things: a survey. *Journal of Industrial Information Integration* 10: 1–9. https://doi.org/10.1016/j.jii.2018.01.005.

**2** Chiang, M. and Zhang, T. (2016). Fog and IoT: an overview of research opportunities. *IEEE Internet of Things Journal* https://doi.org/10.1109/JIOT.2016.2584538.

**3** Akpakwu, G.A., Silva, B.J., Hancke, G.P., and Abu-Mahfouz, A.M. (2017). A survey on 5G networks for the internet of things: communication technologies and challenges. *IEEE Access* https://doi.org/10.1109/ACCESS.2017.2779844.

**4** Cheng, J., Chen, W., Tao, F., and Lin, C.L. (2018). Industrial IoT in 5G environment towards smart manufacturing. *Journal of Industrial Information Integration* https://doi.org/10.1016/j.jii.2018.04.001.

**5** Ficzere, D., Soós, G., Varga, P. et al. (2021). Real-life V2X measurement results for 5G NSA performance on a high-speed motorway. *IFIP/IEEE International Symposium on Integrated Network Management (IM)*, pp. 836–841.

**6** Waring, J. (2018). China to take 40% of 5G connections in 2025. *Mobile World Live.* https://www.mobileworldlive.com/asia/asia-news/china-to-take-40-of-5g-connections-in-202.

**7** Ngo, H.Q. (2021). Massive MIMO. In: *5G and Beyond* (ed. X. Lin and N. Lee). Cham: Springer https://doi.org/10.1007/978-3-030-58197-8_4.

**8** Haque, A.K.M.B. and Bhushan, B. (2021). Security attacks and countermeasures in wireless sensor networks. In: *Integration of WSNs into Internet of Things*, 17–43. https://doi.org/10.1201/9781003107521-2.

**9** Pranto, T.H., Noman, A.A., Mahmud, A., and Haque, A.K.M.B. (2021). Blockchain and smart contract for IoT enabled smart agriculture. *PeerJ Computer Science* 7: https://doi.org/10.7717/peerj-cs.407.

**10** Yang, Z., Yue, Y., Yang, Y. et al. (2011). Study and application on the architecture and key technologies for IOT. *International Conference on Multimedia Technology*, pp. 747–751. https://doi.org/10.1109/ICMT.2011.6002149.

**11** Mistry, S.T., Tyagi, S., and Kumar, N. (2020). Blockchain for 5G-enabled IoT for industrial automation: a systematic review, solutions, and challenges.

*Mechanical Systems and Signal Processing* 135: 106382. https://doi.org/10.1016/j.ymssp.2019.106382.

**12** AKM, B.H. (2019). Need for critical cyber defence, security strategy and privacy policy in Bangladesh—hype or reality? *International Journal of Managing Information Technology* 11 (01): 37–50. https://doi.org/10.5121/ijmit.2019.11103.

**13** Bahalul Haque, A.K.M., Bhushan, B., Nawar, A. et al. (2022). Attacks and countermeasures in IoT based smart healthcare applications. In: *Recent Advances in Internet of Things and Machine Learning. Intelligent Systems Reference Library*, vol. 215 (ed. V.E. Balas, V.K. Solanki, and R. Kumar). Cham: Springer https://doi.org/10.1007/978-3-030-90119-6_6.

**14** Samdanis, K. and Taleb, T. (2020). The road beyond 5G: a vision and insight of the key technologies. *IEEE Network* 34 (2): 135–141.

**15** Ding, J., Nemati, M., Ranaweera, C., and Choi, J. (2020). IoT connectivity technologies and applications: a survey. *IEEE Access* 8: 67646–67673.

**16** Minoli, D. and Occhiogrosso, B. (2019). Practical aspects for the integration of 5G networks and IoT applications in smart cities environments. *Wireless Communications and Mobile Computing.*

**17** Goyal, S., Sharma, N., Kaushik, I. et al. (2021). A green 6G network era: architecture and propitious technologies. *Data Analytics and Management*, pp. 59–75. https://doi.org/10.1007/978-981-15-8335-3_7.

**18** Mani Sekhar, S.R., Nidhi Bhat, G., Vaishnavi, S., and Siddesh, G.M. (2021). Security and privacy in 5G-enabled internet of things: a data analysis perspective. In: *Blockchain for 5G-Enabled IoT* (ed. S. Tanwar). Cham: Springer https://doi.org/10.1007/978-3-030-67490-8_12.

**19** Rong B, Han S, Kadoch M, Chen X, Jara A (2020) Integration of 5G networks and internet of things for future smart city. *Wireless Communications and Mobile Computing.* 2903525, 2 pp. https://doi.org/10.1155/2020/2903525.

**20** Gineste, M., Deleu, T., Cohen, M. et al. (2017). Narrowband IoT service provision to 5G user equipment via a satellite component. *Proceedings of the 2017 IEEE globecom workshops (GC Wkshps)*, Singapore (4–8 December 2017), pp. 1–4.

**21** Fang, X., Wei, T., Feng, W. et al. (2020). 5G embraces satellites for 6G ubiquitous IoT: basic models for integrated satellite terrestrial networks. arXiv 2020. arXiv:2011.03182.

**22** Kumhar, M. and Bhatia, J. (2021). Emerging communication technologies for 5G-enabled internet of things applications. In: *Blockchain for 5G-Enabled IoT* (ed. S. Tanwar). Cham: Springer https://doi.org/10.1007/978-3-030-67490-8_6.

**23** Sinha, P., Jha, V.K., Rai, A.K. et al. (2017). Security vulnerabilities, attacks and countermeasures in wireless sensor networks at various layers of OSI reference model: a survey. *2017 International Conference on Signal Processing and Communication (ICSPC)*. https://doi.org/10.1109/cspc.2017.8305855.

**24** Jaitly, S., Malhotra, H., Bhushan, B. (2017). Security vulnerabilities and countermeasures against jamming attacks in wireless sensor networks: a survey. *2017 International Conference on Computer, Communications and Electronics (Comptelix)*. https://doi.org/10.1109/comptelix.2017.8004033.

**25** Mehbodniya, A., Bhatia, S., Mashat, A. et al. (2022). Proportional fairness based energy efficient routing in wireless sensor network. *Computer Systems Science and Engineering* 41 (3): 1071–1082. https://doi.org/10.32604/csse.2022.021529.

**26** Palanisamy, T., Alghazzawi, D., Bhatia, S. et al. (2022). Improved energy based multi-sensor object detection in wireless sensor networks. *Intelligent Automation & Soft Computing* https://doi.org/10.32604/iasc.2022.023692.

**27** Hai, V.T. (2022). Benefits of 5G millimeter-wave communication in IoT applications. *Dept. Comput. Informat., Bournemouth Univ.*, Bournemouth, UK.

**28** Bhushan, B. and Sahoo, G. (2020). Requirements, protocols, and security challenges in wireless sensor networks: an industrial perspective. In: *Handbook of Computer Networks and Cyber Security*, 683–713. https://doi.org/10.1007/978-3-030-22277-2_27.

**29** Varga, P., Peto, J., Franko, A. et al. (2020). 5G support for industrial IoT applications—challenges, solutions, and research gaps. *Sensors* 20 (3): 828. https://doi.org/10.3390/s20030828.

# 8

# Advancement in Resource Allocation for Future Generation of Communications

*Garima Chopra*[1] *and Suhaib Ahmed*[2]

[1] *Chitkara Institute of Engineering and Technology, Chitkara University, Rajpura, Punjab, India*
[2] *Department of Electronics and Communication Engineering, Model Institute of Engineering and Technology, Jammu, J&K, India*

## 8.1 Introduction

The majority of digital technologies currently depend on wireless communication as their enabling technology. The application domain now includes a wide range of applications in wireless sensor networks (WSNs), Internet of Things (IoTs), eHealth, Vehicle to everything (V2X), unmanned aerial vehicles (UAVs), and body area networks, to mention a few, in addition to traditional cellular communications. Due to the limited spectrum, energy, time, and area available for wireless communication, the problem of resource allocation has received a lot of attention lately.

The digital environment is still growing and changing, and more services and use cases are depending on the availability of up-to-date, trustworthy data. The current wave of wireless communication technology, extending beyond 6G, is designed to provide more throughput, speed, dependability, and pervasive connectivity. The 5G key performance indicator's (KPI) shortcomings are illustrated in [1], along with potential areas for future study. There should be worldwide coverage, encompassing space, air, ground, and sea, with peak data speeds of up to terabits per second and user-experienced data rates exceeding 1 Gbps. Target latencies are less than 1 ms. High location accuracy and scalability of connection density to 108 devices/km$^2$ are provided. The assured reliability (security) target is 99.99, 999%. The network should see an increase in energy efficiency of $100 - 10^9$ bits/J. Furthermore, computing powers, including support for machine learning, are widely accessible on small IoT devices and at the mobile edge. A new proposal [2] including both new and improved capabilities

*Current and Future Cellular Systems: Technologies, Applications, and Challenges*, First Edition.
Edited by Garima Chopra, Suhaib Ahmed, and Shalli Rani.

for International Mobile Telecommunications 2030 (IMT-2030) has been released by the IMT-2030 Promotion Group.

## 8.2 Current Trends in Multiple Access Techniques

The continuous development of cellular networks and the growing need for faster data speeds, reduced latency, and more effective spectrum use have impacted current developments in multiple access strategies in wireless communications. The following are some major trends:

- Orthogonal Frequency Division Multiple Access (OFDMA):
  Widely used in 4G Long-Term Evolution (LTE), OFDMA is still a key component of 5G networks. By dividing the available spectrum into several orthogonal subcarriers, it enables numerous users to receive data simultaneously.
- Non-Orthogonal Multiple Access (NOMA):
  This technique is gaining interest as a potentially useful approach for networks in the future, especially for 5G and beyond. Power domain multiplexing (Power Domain NOMA) and code domain multiplexing (Code Domain NOMA) allow numerous users to share the same time-frequency resources, enhancing spectral efficiency and facilitating huge connectivity.
- Dynamic Spectrum Access (DSA):
  DSA improves spectrum efficiency by providing opportunistic access to underutilised or momentarily available spectrum bands. This includes cognitive radio networks and methodologies such as Licenced Shared Access (LSA).
- Massive MIMO (Multiple Input Multiple Output):
  This technology increases spectral efficiency, improves coverage, and lowers interference by serving several customers at once by utilizing a large number of antennas at the base station.
- Interference Alignment:
  The goal of interference alignment approaches is to maximize system capacity and spectrum efficiency by managing and aligning interference, especially in settings of dense deployment.
- Resource Allocation and Scheduling Algorithms:
  Sophisticated algorithms for these two critical processes are essential for maximizing network performance, guaranteeing user equity, and adjusting to changing traffic and channel circumstances.
- Hybrid Methods of Access:
  Integrating various multiple access methods (for example, OFDMA and NOMA) to make use of their unique benefits and maximize performance in a range of network conditions.

- Machine Learning and AI in Multiple Access:
  The combination of artificial intelligence and machine learning methods to improve network performance, forecast user behavior, manage resources adaptively, and optimize multiple access systems.
- Ultra-Reliable Low Latency Communication (URLLC):
  Techniques tailored for applications requiring ultra-low latency and high reliability, such as industrial IoT and autonomous vehicles, where stringent requirements on latency and reliability are critical. Industry 5.0 revolutionizes work processes by fusing human knowledge with cutting-edge technologies, which is essential for the development of smart cities. It overcomes obstacles including data-intensive analysis and safety issues by using fog computing to improve the efficiency, security, and immediate analysis of Industrial Internet of Things (IIoT) sensors [3].
- Green and Energy-Efficient Techniques:
  Techniques focused on reducing energy consumption and improving the energy efficiency of wireless networks, such as sleep modes for devices i.e. following base station sleeping approaches, dynamic power control, and energy-efficient scheduling algorithms [4].

These trends illustrate the ongoing innovation and adaptation in multiple access techniques to meet the evolving demands of wireless communications, ranging from enhanced spectral efficiency and capacity to lower latency and improved reliability for diverse applications. In the next section, we will discuss the scheduling algorithms in detail including next generation.

Table 8.1 includes details about the latest multiple access schemes, highlighting their strengths, weaknesses, and applications:

These advanced multiple access schemes (such as Index Modulation Multiple Access [IMMA], NOMA, Rate-Splitting Multiple Access [RSMA], etc.) provide increased efficiency, capacity, and flexibility for a range of applications, all while meeting the rigorous specifications of both current and future communication systems.

## 8.3 Scheduling Algorithms for 5G/Beyond 5G

Scheduling algorithms will be essential for 5G and beyond since they optimize network efficiency, efficiently manage resources, and boost overall performance [6]. The following important scheduling algorithms are either in use or have been suggested for 5G and beyond 5G networks:

1) **Proportional Fairness (PF):** While maintaining equity for all users, our algorithm gives preference to those with favourable channel conditions.

**Table 8.1** Comparison of multiple access schemes from conventional to latest schemes highlighting their applications.

| MA schemes | Strength(s) | Weaknesses | Applications |
|---|---|---|---|
| Time division multiple access (TDMA) | Efficient use of bandwidth, simple implementation, easy to manage time slots | Time synchronization required, not suitable for variable-rate data | Cellular networks (2G) |
| Frequency division multiple access (FDMA) | Simple implementation, no time synchronization required | Inefficient use of spectrum, fixed frequency allocation | Traditional analog cellular systems (1G), radio broadcasting |
| Code division multiple access (CDMA) | High capacity, robust against interference and eavesdropping | Complex implementation, requires precise power control | 3G cellular networks, GPS |
| Orthogonal frequency division multiple access (OFDMA) | High spectral efficiency, robust to multipath fading, flexible allocation of resources | High computational complexity, sensitive to frequency offset | 4G/5G cellular networks, Wi-Fi (802.11ax) |
| Space division multiple access (SDMA) | Increases capacity by exploiting spatial separation | Requires advanced antenna technology and processing | MIMO systems, satellite communications |
| Carrier sense multiple access with collision avoidance (CSMA/CA) | Simple and decentralized, effective for small networks | Collisions still possible, not suitable for high traffic | Wi-Fi (802.11), ethernet networks |
| Carrier sense multiple access with collision detection (CSMA/CD) | Efficient for low to medium traffic, simple implementation | Performance degrades with high traffic, not suitable for wireless | Ethernet (wired LANs) |

| | | | |
|---|---|---|---|
| Non-orthogonal multiple access (NOMA) | High spectral efficiency, supports massive connectivity, improved fairness | Increased receiver complexity, power control challenges | 5G networks, IoT, vehicular communications |
| Sparse code multiple access (SCMA) | High spectral efficiency, supports massive connectivity, low latency | Complex receiver design, computationally intensive | 5G networks, machine-type communications, IoT |
| Grant-free multiple access (GFMA) | Low latency, supports massive connectivity, efficient for sporadic traffic | Limited to low data rates, potential for collisions | IoT, machine-type communications, sensor networks |
| Rate-splitting multiple access (RSMA) | Efficiently manages interference, needed for broadcasting information | Receiver complexity due to additional Successive Interference cancellation (SIC) layer | 5G/beyond 5G communication |
| Index modulation multiple access (IMMA) [5] | Increased spectral efficiency, robustness to interference, low hardware complexity | Limited by index modulation constraints, not widely adopted yet | 6G wireless communications, IoT |

By assigning resources to customers according to a ratio of their data rate to their average channel quality, it seeks to maximise throughput overall.

2) **Maximal Ratio Combining (MRC):** MRC maximises the signal-to-noise ratio (SNR) by scheduling transmissions depending on the ratio of received signal intensities from numerous antennas. It includes increased diversity gain and enhanced dependability in fading channels. It is especially helpful for increasing spectral efficiency in MIMO systems.
3) **Weighted Round Robin (WRR):** This method of resource allocation is cyclical and gives priority to users who have greater weights among various traffic kinds or users. It includes the ability to prioritise different forms of traffic (such as speech over data) and the flexibility to allocate resources. It is beneficial in situations where distinct Quality of Service (QoS) guarantees are needed for various types of traffic.
4) **QoS Scheduling Algorithm:** QoS-aware scheduling assigns customers a priority according to their QoS needs, including dependability, latency, and jitter. QoS Scheduling algorithms enhances user data rate requirement and ensures QoS guarantees for important applications. It is crucial for industrial IoT, telemedicine, and real-time video streaming applications. DSA, or dynamic spectrum access: DSA automatically adjusts allocations based on actual spectrum availability, enabling users to take advantage of underutilised spectrum bands. **Benefits:** Addresses spectrum shortages and improves spectrum utilisation efficiency. Use Cases: Especially pertinent in situations with variable spectrum availability and in environments involving spectrum sharing. Scheduling Determined by Machine Learning (ML): **Description:** ML algorithms use current network circumstances and previous data to dynamically optimise resource allocation. Benefits: Adaptable to shifting network conditions, with the potential to beat conventional algorithms in challenging settings. Use Cases: becoming more and more of a viable strategy in B5G networks because of its versatility and capacity to optimise several parameters at once.

## 8.4 Factors Influencing Scheduling Algorithms

The network frequently uses the channel state information (CSI) reports that the devices produce to inform scheduling decisions. The scheduling algorithm generates a metric to ascertain when the user can access the channel by gathering several metrics from reports provided by the user equipment [7, 8]. This subsection describes the scheduling inputs and demonstrates how these factors impact the scheduling algorithms' outputs.

- The performance, efficiency, and reliability of the network are optimised by a number of factors that impact scheduling decisions in wireless communication. Requirements for QoS are critical, including latency, which guarantees timely data delivery for applications that are sensitive to delays, such as VoIP and video streaming. Reliability ensures reliable data transfer, while throughput is essential for high data rate applications. For applications like live streaming that require a constant data flow, jitter minimization is crucial.
- Decisions on scheduling are greatly influenced by channel conditions. Better SNR customers are thought to be prioritised, guaranteeing higher data rates and effective spectrum utilisation. In order to guarantee adequate communication quality for every user, scheduling must also be adjusted to prevent fading effects, maintain fair access even in weak signal situations, and manage interference levels.
- User mobility is still another important element. Moving continuously affects scheduling in order to keep dependable connections, particularly when handovers for quickly moving users are involved. Resource allocation to maximize network performance is also influenced by the user's location within the cell, whether it is at the edge of the centre.
- To avoid congestion and guarantee efficient network functioning, traffic load balancing is essential. By examining past, present, and anticipated traffic patterns, scheduling can be proactively adjusted to effectively meet changing demands. Allocating resources fairly guarantees that users are distributed equally, preventing any one user from controlling resources and preserving user satisfaction levels overall. Maintaining equitable access for all users while optimising overall network throughput is the goal of proportional fairness [9].
- Energy efficiency is becoming more and more crucial, especially for IoT networks and mobile devices. In order to balance energy consumption and communication quality, scheduling can control transmission power and give priority to mobile device battery conservation. To maintain optimal performance, admission control controls network access based on regulations and current conditions.
- Scheduling considerations are also influenced by application needs. Different applications demand different resource allocations, such as real-time and non-real-time services. Respecting Service Level Agreements (SLAs) sets scheduling priority by guaranteeing performance for particular users or services. To adapt to changing demands and circumstances, the availability of resources, including bandwidth and spectrum allotment, must be dynamically controlled.
- Technology limitations need to be taken into account, such as the capabilities and limitations of user devices and network hardware. To ensure interoperability and compliance with legal requirements, scheduling decisions are also

influenced by adherence with communication requirements and protocols. Adaptive scheduling is required to maintain communication quality due to environmental constraints such as physical impediments in metropolitan contexts and meteorological conditions which lead to signal degradation (e.g., interruption in communication through satellite due to bad weather conditions).

- Finally, predictive analysis and historical data are important. Optimized scheduling is possible by the prediction of future usage through the analysis of usage trends from historical data. By putting machine learning algorithms into practice, one can improve overall network performance by enhancing predictive skills and enabling dynamic scheduling modifications.

To sum up, scheduling in wireless communication involves striking a balance between several variables to optimise network effectiveness, efficiency, and user happiness. To efficiently respond to changing conditions and requirements, sophisticated algorithms and real-time modifications are required.

## 8.5 Resource Allocation for 5G Ultra-Dense Networks

HetNets have been seen as a potential paradigm to handle the deluge of data traffic since they offer end users quicker uplink and downlink data rates, decreased latency, and wider bandwidths [10]. HetNets are made up of several tiny cells nested within standard macro cells. Today's HetNets are becoming increasingly densely packed, known as ultradense HetNets. To meet the need for the capacity of the upcoming 5G wireless communication network to expand 1000 times. HetNets are made up of ultra-dense networks comprising many different access points (APs) that are densified to fulfil high-capacity demands in a range of situations. These include low-power small cells (such as pico, femto, and microcells), traditional high-power macro cells, and other noncellular heterogeneous communication systems (such as Wireless local Area Network (WLAN), direct-to-direct [D2D], Low-Power Wide-Area Network [LPWAN], etc.). Ultra-dense HetNets would benefit from significantly increased spatial reuse of the limited frequency resources due to the dense deployment of heterogeneous nodes. They would also offer greater installation flexibility to improve coverage and capacity, particularly for hotspot and blind wireless areas [11]. Due to extensive benefits of dense networks in 5G, still it suffers from challenging issues as a result of dynamic network architecture for different applications. In this section, we discuss the RA schemes in Ultra-dense networks for 5G communication.

- Resource Management in ultra-dense network (UDN)
  The past decade has seen a proliferation of new services, including computation-intensive, content-centric, delay-sensitive, and ultra-reliable services, due to the

popularity of mobile devices and the exponential increase of mobile Internet traffic. As a result, manageable resources in wireless communication systems are changing from being solely wireless resources—such as power, spectrum, time, and space—to being comprehensive resources that include computational capacity, renewable energy sources, backhaul and fronthaul capacity, battery resources, storage availability, etc. Upgrade difficulties arise for the RA problems and corresponding allocation schemes given different available resources, which are always exhibited by varying combinations (spectrum & power, battery resources & computation capability & storage availability, etc.).

- QoS Assurance
  QoS requirements are becoming increasingly varied and stringent due to the rapid expansion of innovative applications such as VR/AR, smart transportation, drones, robotics, online gaming, online video, online reading, and e-health monitoring [12]. Therefore, these particular applications—such as ultra-low latency, ultra-high dependability, increased data rate, computation efficiency, or even combinations of them—should be taken into consideration while designing RA schemes for networks. Simultaneously, the network may undergo extensive modifications to satisfy the escalating QoS demands. For example, edge computing, also known as fog computing, and edge caching are generally accepted to be essential additions that help to reduce the disparity between the required QoS and the available wireless capacity [13–17].
- Fairness Guaranteed
  Fairness and priority consideration are also crucial during the RA process in UDNs. Fair Resource Allocation (RA) systems guarantee that each AP/user receives an equitable proportion of resources, rather than focusing on the overall system or average level performance. Max-min fairness [18], proportional fairness [19, 20], weighted proportional fairness [21], adaptive proportional fairness [22], alpha-fairness [23], etc. are a few of the fairness principles that are frequently applied in resource allocation situations. Two types of fairness can be distinguished based on the duration of the observation period: long-term fairness and short-term fairness. Conversely, priority guarantees that the individual resources are enjoyed by the AP/user in accordance with their priority.
- Coordinating such a large number of APs/users is computationally demanding as network densification becomes an inescapable pattern. Several studies have focused on effective random assignment strategies to lower computing complexity [19, 20, 24]. The RA process can be broken down into multiple smaller problems or carried out in a distributed manner by utilising techniques such as convex optimization, game theory, graph theory, stochastic geometry methods, stochastic optimization, grouping/clustering method, etc. This will greatly reduce the computational complexity. Recently, [5] explores into the

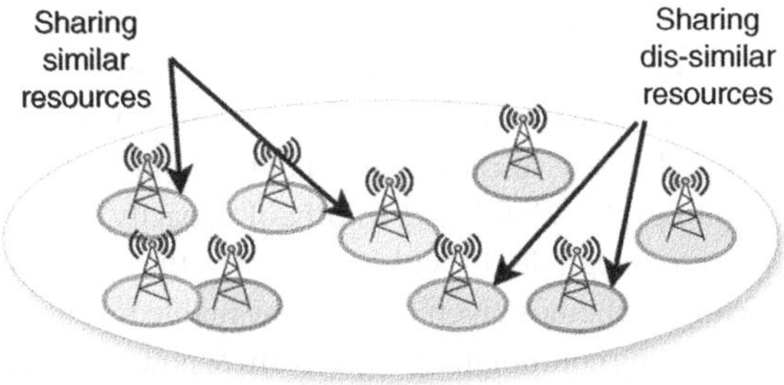


**Figure 8.1** Resource sharing through mutual cooperation.

best decomposition for large-scale infrastructure-based wireless networks from a graph-theoretic perspective. This approach offers a mean of RA and interference control by dividing the large-scale network into multiple manageable tiny sub-networks.

- Mutual Cooperation: In Figure 8.1, the dense deployment of APs makes the assignment of resources a difficult task with each AP having different load conditions. Therefore, depending on the load conditions, all the BSs mutually cooperate among themselves by forming groups. Each group consists of few BSs sharing similar resources, thereby, reducing the interference among them and the users associated with each cell. This approach is beneficial in the sense that it leads to efficient utilization of resources. The BSs with less load are engaged with the heavily loaded BSs. This overlapping approach handles the pressure of BSs abundance density by efficiently managing them.

## 8.6 Conclusion

Multiple access technologies, which control how radio resources are distributed among users, have a basic impact on transceiver complexity and system performance, and their development is strongly linked to the evolution of wireless cellular networks. In multiple access schemes, an efficient design for resource allocation is crucial because it guarantees effective management of radio resources, reduces interference, and offers dependable, fast communication services to a large number of users. With an emphasis on the context-specific nature of resource allocation design and the variety of optimisation methods used in NGMA systems, this chapter has presented a thorough examination of the subject.

## References

**1** Wang, C.-X., You, X., Gao, X. et al. (2023). On the road to 6G: Visions, requirements, key technologies, and testbeds. *IEEE Communications Surveys and Tutorials* 25 (2): 905–974.

2 (2022). Future technology trends of terrestrial international mobile telecommunications systems towards 2030 and beyond, document ITU M.2516, ITU.

3 Shruti, S.R. and Srivastava, G. (2024). Secure hierarchical fog computing-based architecture for industry 5.0 using an attribute-based encryption scheme. *Expert Systems with Applications* 235: 121180.

4 He, Z., Mao, S., Kompella, S., and Swami, A. (2015). Minimum time length scheduling under blockage and interference in multi-hop mmWave networks. *2015 IEEE Global Communications Conference (GLOBECOM)*, (December 2015), San Diego, CA, USA, pp. 1–7. IEEE.

5 Li, J., Dang, S., Wen, M. et al. (2023). Index modulation multiple access for 6G communications: principles, applications, and challenges. *IEEE Network* 37 (1): 52–60.

6 Gómez-Cuba, F. and Zorzi, M. (2020). Optimal link scheduling in millimeter wave multi-hop networks with MU-MIMO radios. *IEEE Transactions on Wireless Communications* 19 (3): 1839–1854. IEEE.

7 Castañeda, E., Silva, A., Gameiro, A., and Kountouris, M. (2017). An overview on resource allocation techniques for multi-user MIMO systems. *IEEE Communications Surveys and Tutorials* 19 (1): 239–284. IEEE.

8 Barb, G.-R. and Otesteanu, M. (2018). 5G: an overview on challenges and key solutions. *Proceedings of International Symposium on Electronics and Telecommunications (ISETC)*, (November 2018), pp. 1–4. IEEE.

9 Fan, B., Ramirez, D., Huang, L. et al. (2018). A cross-tier scheduling scheme for multi-tier millimeter wave wireless networks. *IEEE Transactions on Wireless Communications* 17 (8): 5029–5044.

10 Zhang, T., Zhao, J., An, L. et al. (2016). Energy efficiency of base station deployment in ultra dense HetNets: a stochastic geometry analysis. *IEEE Wireless Communications Letters* 5 (2): 184–187.

11 Liu, H., Xu, S. and Kwak, K.S. (2016). Geometric programming based distributed resource allocation in ultra dense Hetnets, *Proceedings of IEEE Vehicular Technology Conference*, Nanjing (May 2016), pp. 1–5.

12 Dogra, R., Rani, S., Sharma, B., and Verma, S. (2021). Essence of scalability in wireless sensor network for smart city applications. *IOP Conference Series: Materials Science and Engineering*, vol. 1022, no. 1, p. 012094, IOP Publishing. 1st International Conference on Computational Research and Data Analytics (ICCRDA 2020) 24th October 2020, Rajpura, India.

13 Sardellitti, S., Scutari, G., and Barbarossa, S. (2015). Joint optimization of radio and computational resources for multicell mobile-edge computing. *IEEE Transactions on Signal and Information Processing over Networks* 1 (2): 89–103. IEEE.

14 Mao, Y., Zhang, J., and Letaief, K.B. (2017). Joint task offloading scheduling and transmit power allocation for mobile-edge computing systems. *Proceedings*

*of IEEE Wireless Communications and Networking Conference (WCNC)*, San Francisco, CA (March 2017), pp. 1–6.

**15** Yu, Y., Zhang, J., and Letaief, K.B. (2016). Joint subcarrier and CPU time allocation for mobile edge computing. *Proceedings of IEEE Global Communications Conference (GLOBECOM)*, Washington, DC (December 2016), pp. 1–6. IEEE.

**16** Wang, F., Xu, J., Wang, X. et al. (2017). Joint offloading and computing optimization in wireless powered mobile-edge computing systems. *Proceedings of IEEE International Conference on Communications (ICC)*, Paris (May 2017), pp. 1–6. IEEE.

**17** Mao, Y., Zhang, J., and Letaief, K.B. (2016). Dynamic computation offloading for mobile-edge computing with energy harvesting devices. *IEEE Journal on Selected Areas in Communications* 34 (12): 3590–3605.

**18** Liu, A. and Lau, V. (2014). Phase only RF precoding for massive MIMO systems with limited RF chains. *IEEE Transactions on Signal Processing* 62 (17): 4505–4515. IEEE.

**19** Xu, J., Wang, J., Zhu, Y. et al. (2014). Cooperative distributed optimization for the hyper-dense small cell deployment. *IEEE Communications Magazine* 52 (5): 61–67.

**20** Chuang, M.C., Chen, M.C., and Lin, Y.H. (2016). SDN-based resource allocation scheme in ultra-dense OFDMA smallcell networks. *Proceedings of International Conference on Advanced Materials for Science and Engineering (ICAMSE)*, Tainan (November 2016), pp. 524–527.

**21** Hamda, S., Pischella, M., Roviras, D., and R. Bouallegue (2013). Analysis of weighted proportional fair resource allocation for uplink OFDMA. *Proceedings of IEEE 78th Vehicular Technology Conference (VTC Fall)*, Las Vegas, NV (September 2013), pp. 1–5.

**22** Wang, S., Huang, F., and Wang, C. (2013). Adaptive proportional fairness resource allocation for OFDM-based cognitive radio networks. *Wireless Networks* 19 (3): 273–284.

**23** Xu, P., Cumanan, K., and Yang, Z. (2017). Optimal power allocation scheme for NOMA with adaptive rates and alpha-Fairness. *Proceedings of 2017 IEEE Global Communications Conference (GLOBECOM)*, Singapore (December 2017), pp. 1–6.

**24** Awad, J.F., Imran, M.A., and Tafazolli, R. (2013). Reduced complexity interference avoidance scheme based on user grouping in dense cellular networks. *Proceedings of 19th European Wireless Conference*, Guildford, UK (April 2013), pp. 1–6.

# 9

# Next-Gen Networked Healthcare: Requirements and Challenges

*Kanica Sachdev[1] and Brejesh Lall[1]*

[1] *Bharti School of Telecommunication Technology and Management, IIT Delhi, New Delhi, India*

## 9.1 Introduction

Telemedicine, a concept rooted in the mid-nineteenth century, has evolved significantly, catalyzed by technological advancements [1]. From the telegraph's use during the Civil War to transmit messages about wounded soldiers, to William Einthoven's pioneering electrocardiogram transmission in 1905, telemedicine has continually pushed the boundaries of healthcare accessibility [1, 2]. The amalgamation of technology and healthcare, particularly with the proliferation of smartphones, has transformed the landscape, enabling real-time examinations and fostering innovative applications. The recent convergence of fifth-generation wireless networks (5G), artificial intelligence (AI), the Internet of Things (IoT), and blockchain has sparked a paradigm shift in telemedicine [3]. The onset of the SARS-CoV-2 pandemic in Wuhan acted as a catalyst, propelling the adoption of 5G technology to address real-time healthcare needs, a feat unattainable with the limitations of 4G networks [4, 5]. The pandemic underscored the pivotal role of telemedicine in delivering medical assistance globally, emphasizing its efficacy in triaging patients and minimizing unnecessary exposure risks [6–8]. As of May 4, 2022, the World Health Organization reported staggering figures of 515 million COVID-19 cases and over 6 million deaths worldwide, highlighting the urgency of leveraging technology to address healthcare challenges [9]. The pandemic marked the early global emergency use of 5G technology, demonstrating its vital role in providing seamless communication across various sectors, particularly healthcare, where 4G networks fell short [9].

The introduction of 5G networks extends beyond connecting humans; it encompasses a vast ecosystem of smart devices [10]. Projections estimate 25 billion IoT devices by 2025 and a staggering 600 billion by 2030. This proliferation enables

*Current and Future Cellular Systems: Technologies, Applications, and Challenges*, First Edition.
Edited by Garima Chopra, Suhaib Ahmed, and Shalli Rani.

continuous data acquisition and monitoring from machines, equipment, and living systems, ushering in an era of interconnected healthcare [11]. Integrating 5G into healthcare not only promises unprecedented speed, ultra-low latency, and massive device connectivity but also addresses reliability concerns, fostering transformative applications.

The integration of 5G into healthcare offers a myriad of transformative applications. From remote robotic-assisted surgery and wearable devices to medical big data management, the capabilities of 5G hold promise in revolutionizing healthcare delivery [12–15]. However, the diverse communication technical requirements of different healthcare use cases pose a challenge. Stakeholders, including developers, network providers, and regulatory authorities, need a nuanced understanding of these requirements to design, deploy, and regulate 5G-enabled healthcare applications effectively [13]. Quantitative key performance indicators (KPIs) become crucial in assessing the feasibility of these use cases, ensuring the required level of service for safe and effective healthcare applications [16].

Despite the immense potential, deploying 5G in healthcare demands a cautious approach. Critical biological and physiological factors must be considered, especially with regard to leveraging the full spectrum of 5G capabilities. While 5G promises low latency, high bandwidth, and reduced energy consumption, signal degradation at higher frequencies poses challenges, necessitating densely populated "base stations" [11]. Furthermore, considerations of human interface technology, spectrum exploitation challenges, and the interplay of various technologies such as AI and blockchain in the 5G ecosystem demand careful investigation.

In this paper, we delve into the vast landscape of 5G technology and its transformative potential within the healthcare sector. Our exploration encompasses a summary of the applications that 5G introduces to healthcare. We also analyze the technological prerequisites essential for the seamless integration of 5G into healthcare systems, emphasizing the importance of understanding diverse communication requirements for different healthcare use cases. Stakeholders, including developers, network providers, and regulatory authorities, must navigate these technical requirements adeptly to deploy and regulate 5G-enabled healthcare applications effectively. We also outline the challenges that accompany this technological evolution, and the critical considerations of human interface technology, spectrum exploitation, and the intricate interplay of various technologies within the 5G ecosystem.

## 9.2 Applications

In the realm of healthcare, the advent of 5G technology heralds a transformative era, unlocking a spectrum of applications poised to reshape medical practices. With its ultra-low latency and high bandwidth, this technology propels advancements in wearable health devices, facilitates emergency response, and revolutionizes medical training through augmented reality. The diverse applications of 5G in healthcare hold the promise of enhancing patient care, expanding accessibility, and ushering in a new paradigm of efficient and connected healthcare systems. Here are some key areas where 5G can make a significant impact:

### 9.2.1 Remote Robotic-Assisted Surgery

Remote Robotic-Assisted Surgery, a pinnacle of 5G-enabled healthcare, revolutionizes the landscape of surgery. Through remote care, healthcare specialists transcend geographical barriers to provide medical services via the Internet, sharing vital physiological data collected by sensors and the Internet of Medical Things (IoMT) devices [17]. The fusion of advanced communication technologies and medical expertise opens avenues for enhanced patient care, positioning remote robotic-assisted surgery as a cornerstone in the evolution of healthcare. This paradigm shift is exemplified in various implementations, such as remote laparoscopic surgery and retinal photocoagulation for diabetic retinopathy.

In a groundbreaking illustration, Zheng et al. implemented a remote laparoscopic surgery system, where a surgical robot named "Micro Hand" was manipulated by a surgeon located 3000 km away [18]. The successful surgery on a pig showcased the potential of this technology in overcoming geographical constraints. Similarly, Chen et al. reported the development of a system for retinal photocoagulation in diabetic retinopathy patients, marking the first remotely controlled procedure of its kind [19]. Gupta et al. introduced a telesurgery architecture, utilizing a master–slave network to remotely manipulate a surgical arm [20]. Meshram and Patil proposed a tactile robotic telesurgery architecture, enhancing surgical precision through direct and feedback links [21]. The implications of these technologies extend to the realm of disease diagnosis, with models like Deep Belief Network (DBN) proving instrumental in analyzing and classifying medical data.

### 9.2.2 Remote Diagnosis and Teleconsultation

Through advanced telemedical applications, patients can benefit from timely diagnostic services and consultations without the need for physical presence. The implementation of remote diagnosis systems has been exemplified by Zhang et al., who presented a remote clinic and ambulance system at Zhengzhou University Hospital [22]. This system allows patients to undergo diagnosis and treatment without the necessity of traveling to the hospital. High-definition video consultations have been successfully conducted, illustrating the potential for efficient remote healthcare services. In the realm of ophthalmology, Chen et al. reported the development of a system for retinal photocoagulation in diabetic retinopathy patients [19]. This teleophthalmology platform integrates a laser device, a remote computer control platform, and a teleconsultation platform through video conferencing. It marks a significant milestone as the first remotely controlled retinal photocoagulation procedure. Wang et al. demonstrated the application of a remote ultrasound system for diagnosing COVID-19, allowing comprehensive examinations of the heart, lungs, blood vessels, and inferior vena cava [23]. The innovative system comprises a dual-environment setup, with the doctor stationed in the first environment, equipped with a workstation, console, image monitors, and a transducer. The patient, situated in the second environment, is provided with an ultrasound system and a robotic arm for seamless examinations. In the critical care context, Duan et al. proposed a teleultrasound diagnostic system for intensive care units (ICUs) [24]. The system utilizes ultrasound assisted by a robotic arm, offering a comprehensive examination of vital organs. The incorporation of audio, video, and mainframe systems on both the surgeon's and patient's sides ensures effective communication and diagnostic procedures. The application of 5G in remote diagnosis extends further to specialized areas like postpartum hemorrhage diagnosis. Zhang et al. developed a diagnostic system employing machine learning algorithms such as support vector machines (SVM) and artificial neural networks (ANN) [25]. This model, trained on a dataset of 3842 vaginal delivery cases, showcases the potential for accurate diagnostic predictions.

The 5G-enabled infrastructure not only ensures real-time communication but also opens avenues for predictive and preventive healthcare. Ye emphasized the use of remote systems during the COVID-19 pandemic, incorporating AI for prediction and detection. X-ray images and learning algorithms assist specialists in making informed decisions, showcasing the potential of these technologies in crisis management.

### 9.2.3 In-Ambulance Treatment

In-Ambulance Treatment by Remote Physicians represents a groundbreaking advancement in emergency medical services, leveraging the capabilities of 5G technology to redefine pre-hospital care and enhance patient outcomes. The real-time transmission of critical information, including video, images, and vital signs, within ambulances becomes paramount in emergencies, ensuring that individuals at risk receive prompt and effective care when their health is compromised. As emphasized by Usman et al. [26], the transmission of vital signs, such as oxygen saturation level, heart rate, body temperature, and more, coupled with video transmission, enables remote physicians to observe and support paramedics attending to emergency situations. This not only facilitates real-time diagnosis but also allows doctors to remotely examine ultrasound (US) images acquired in the ambulance, aiding in generating accurate diagnoses and determining appropriate treatment plans.

Zhang et al.'s mobile diagnostic and treatment system, incorporating a remote clinic and ambulance, exemplifies the potential of 5G in transforming ambulances into mobile healthcare units [22]. The proposed smart ambulance architecture by Zhai et al. further underscores the healthcare benefits of 5G-powered in-ambulance treatment [27]. Equipped with a multi-monitoring system, including vital-signs-acquisition devices, a medical workstation, and an HD camera, the smart ambulance facilitates audio and video transmission to hospitals. The use of virtual reality (VR) glasses enhances the remote video transmission experience, allowing doctors to feel immersed and connected during the transfer process. In the broader healthcare landscape, the benefits of In-Ambulance Treatment by Remote Physicians extend beyond emergency situations. The seamless relay of patients' vitals in real time, facilitated by 5G-powered sensors and mobile health monitoring systems, ensures constant and immediate attention through intelligent systems. This category represents a significant portion of clinical applications, constituting 52.27% of the identified benefits.

### 9.2.4 Remote Patient Monitoring

Remote Patient Monitoring: Facilitated by the capabilities of 5G technology, remote patient monitoring represents a significant stride in healthcare, particularly benefiting patients with chronic conditions. The seamless integration of high-bandwidth 5G mobile networks with advanced sensors and wearables creates a powerful ecosystem [28]. It encompasses a diverse range of sensors that can detect various health parameters, including Electro Cardio Graph (ECG) signals, blood pressure, body temperature, glucose levels, and more. These sensors, integrated into 5G-powered wearables and other remote monitoring devices, offer a comprehensive and real-time view of a patient's health metrics. One of the transformative aspects in remote patient monitoring is its potential to enhance drug intake monitoring. Ingestible sensors, a notable advancement in this realm, enable the remote assessment of patient compliance with pharmacotherapy. This capability holds promise in ensuring that patients adhere to prescribed medication regimens, thereby contributing to the overall effectiveness of treatment plans. Such connectivity ensures proactive interventions and personalized care plans tailored to individual patient needs.

The significance of a high-bandwidth 5G network becomes evident in the context of remote patient monitoring. With a large number of IoMT devices in use, the need to share substantial amounts of data remotely necessitates the capabilities provided by 5G. This not only ensures the swift transmission of data but also contributes to the seamless integration of wearables and sensors into the healthcare ecosystem.

### 9.2.5 Medical Big Data Management

Bolstered by the capabilities of 5G technology, medical big data management spearheads a paradigm shift in healthcare [29]. The convergence of massive healthcare data generated by diverse devices and systems finds a robust ally in the efficiency and speed of 5G, revolutionizing medical research, treatment personalization, and the development of predictive healthcare models. Leveraging cloud processing with 5G minimizes the energy needed for data computation on IoMT devices, ensuring prolonged operational efficiency. The high bandwidth of 5G enables seamless transfer of large files across the globe for real-time remote processing, facilitating researchers' access to data for scientific studies and enhanced data analysis, especially pertinent in the era of burgeoning AI applications in medical diagnostics.

Several innovative models and approaches exemplify the synergy of 5G and medical big data management. Low Cost Health Monitoring (LCHM) methods utilize sensors for real-time ECG monitoring, albeit with limitations in response time [30]. FogCepCare, an IoT-relied clinical management method, integrates cloud and sensor layers to expedite health condition identification [31]. The integration of Software Defined Network (SDN) with IoT e-health services presents a novel approach for diagnosing heart attacks through mobile apps [32]. Smart Fog Gateway (SFG) methods focus on grouping data conditioning tasks, enhancing performance optimization in wearable IoT tools [33]. Fog-based Efficient Manufacture Inspection (FEMI) and Fog-based IoT-Healthcare (FIH) solutions demonstrate the efficacy of fog computing platforms in data processing [34, 35]. FogLearn, a fog-based machine learning method, predicts patients prone to Diabetes Mellitus through real-time feature data [36]. Scalable and Accurate Deep Learning (SADL) approaches utilizing Fast Healthcare Interoperability Resources (FHIR) format showcase the potential for accurate clinical action predictions [37]. Service Level Agreement (SLA)-relied Healthcare Big Data Analytics (SLA-HBDA) structures, incorporating effective data sharing models, outperform traditional models in computational efficiency, emphasizing the pivotal role of 5G in advancing healthcare analytics [38]. The synergy between 5G and the management of extensive medical data is set to reshape the healthcare environment, leading to an era of enhanced patient outcomes.

### 9.2.6 Augmented Reality (AR) and Virtual Reality (VR)

The integration of 5G into medical training not only facilitates real-time data transmission but also ensures a level of reliability that is paramount in critical situations. AR glasses, empowered by 5G, play a pivotal role in supporting surgeons during various medical procedures. These high-tech glasses provide crucial information about the patient's disease condition, offering valuable insights that guide surgeons throughout the surgical process. From real-time data feeds to intricate details relevant to the surgery, AR glasses enhance the surgeon's situational awareness and decision-making, contributing to improved surgical outcomes. Moreover, 5G amplifies the sophistication of haptic/tactile internet in medical training. Haptic technology involves simulating the sense of touch, allowing healthcare professionals to experience a realistic touch sensation in virtual environments. With the ultra-low latency and high reliability of 5G, the haptic feedback can be delivered in real-time, creating a more immersive and responsive training environment. This is particularly valuable in procedures where tactile sensitivity is crucial, such as surgery and other medical interventions.

VR also significantly benefits the capabilities of 5G helthcare, especially in the context of training healthcare professionals through simulated surgeries. Simulated reality, enabled by VR, allows practitioners to immerse themselves in lifelike medical scenarios, honing their skills in a risk-free environment. The 5G network's high data transfer rates, exceeding 2.1 Gbps for a typical 360° VR video stream, ensure smooth and high-fidelity simulations. This ensures that medical professionals receive realistic training experiences, preparıng them for the complexities of real-world healthcare scenarios [39].

### 9.2.7 Emergency Response Strategies

5G technology also play a pivotal role in revolutionizing emergency response strategies by enabling swift and efficient communication during crises, natural disasters, or emergencies. This innovative application ensures that emergency responders can coordinate their efforts seamlessly, share real-time data, and deploy resources efficiently to areas affected by the unfolding events. 5G networks provide ultra-fast and low-latency connectivity, allowing emergency responders to establish quick and resilient communication channels. 5G's capabilities empower these responders to communicate seamlessly with each other, irrespective of their locations. Whether it's firefighters, paramedics, law enforcement, or other relevant agencies, the real-time connectivity offered by 5G enhances their ability to collaborate and share crucial information. Additionally, the IoT devices, powered by 5G, can be deployed in emergency scenarios to provide valuable data. These devices, such as sensors and cameras, can offer insights into environmental conditions, structural integrity, and other critical factors. 5G-enabled emergency response transforms the way responders handle crises by providing rapid, reliable, and collaborative communication. As societies continue to face various challenges, the integration of 5G technology into emergency response systems emerges as a crucial step towards building more resilient and responsive communities.

## 9.3 Technological Prerequisites

The integration of 5G into healthcare requires careful consideration of various technological prerequisites to ensure optimal performance, reliability, and security. Several key factors play a crucial role in paving the way for the successful incorporation of 5G technology into healthcare systems. Here are the technological prerequisites essential for this integration [40, 41]:

- **Network Infrastructure:** At the core of successful 5G integration is a resilient network infrastructure characterized by high-speed connectivity. This involves the strategic deployment of small cells, base stations, and other essential network elements. The aim is to establish a pervasive network that not only offers comprehensive coverage but also facilitates low-latency communication. This robust foundation is pivotal for harnessing the full potential of 5G technology in enabling swift and seamless data transmission within the healthcare ecosystem.
- **Edge Computing:** Edge computing plays a pivotal role in enhancing healthcare applications that require minimal latency. It involves the decentralized processing of data, strategically moving computational tasks closer to the data source. In the context of healthcare, this often means situating processing capabilities near medical devices, sensors, or other data-generating endpoints. By doing so, healthcare providers can significantly reduce the time it takes to process and analyze data. This proximity enables faster response times, which is particularly crucial for real-time applications within the healthcare domain. The adoption of edge computing in healthcare ensures that critical data processing occurs swiftly and efficiently, contributing to improved patient care and enabling timely decision-making by healthcare professionals.
- **Security Protocols:** In light of the sensitive nature of healthcare data, the implementation of robust security measures is paramount. End-to-end encryption serves as a cornerstone in safeguarding patient information across its entire journey, encompassing both transmission and storage phases. This comprehensive security protocol significantly mitigates the risk of unauthorized access, data breaches, or potential compromises. By encrypting data at its source and decrypting it only at its intended destination, end-to-end encryption provides an additional layer of defense against cyber threats. This imperative security measure not only ensures the confidentiality and integrity of healthcare information but also instills confidence in patients and healthcare providers alike regarding the privacy and protection of sensitive medical data.

- **Interoperability Standards:** Health Information Exchange (HIE) emerges as a pivotal mechanism for fostering seamless communication and facilitating data sharing across diverse healthcare systems. Its significance lies in adhering to interoperability standards, which are crucial for ensuring compatibility and harmonious interaction among disparate systems. By embracing HIE, healthcare entities can transcend traditional silos, enabling efficient and secure exchange of patient information. This interconnectedness proves instrumental in delivering comprehensive and coordinated care, as healthcare providers gain access to pertinent patient data from various sources. The adoption of HIE not only enhances the overall efficiency of healthcare systems but also contributes to improved patient outcomes through collaborative and informed decision-making.
- **Device Compatibility:** The incorporation of the IoMT devices into healthcare networks necessitates a foundational emphasis on compatibility with 5G technology. Achieving optimal IoMT integration involves meticulous consideration of communication protocols and network architectures to harmonize the diverse ecosystem of healthcare devices with the transformative potential of 5G. This compatibility ensures that IoMT devices can efficiently communicate, share data, and access the high-speed, low-latency features characteristic of 5G networks. By fostering this compatibility, the healthcare landscape stands to benefit from enhanced connectivity, real-time data transmission, and improved responsiveness.
- **Quality of Service (QoS) Management:** Quality of Service (QoS) management plays a pivotal role in the realm of healthcare applications where certain tasks demand prioritized network access. This involves the implementation of sophisticated mechanisms to ensure that critical healthcare services, such as remote surgery or real-time monitoring, are accorded the highest priority in resource allocation within the network. This proactive approach to network resource allocation not only enhances the responsiveness of healthcare systems but also establishes a robust foundation for the successful execution of time-sensitive tasks.
- **Data Management and Storage:** In the realm of data management and storage for healthcare, the integration of cloud infrastructure emerges as a transformative strategy. Leveraging cloud services enhances the scalability and accessibility of healthcare data within 5G-enabled systems. By utilizing the capabilities of the cloud, healthcare organizations can optimize data processing and storage, fostering efficiency and innovation. Furthermore, the accessibility offered by cloud services facilitates seamless data retrieval and analysis, enabling healthcare professionals to derive actionable insights promptly. This integration not only streamlines administrative processes but also lays the foundation for advanced analytics, machine learning, and collaborative research initiatives.

- **Regulatory Compliance:** In the complex landscape of healthcare technology, prioritizing regulatory compliance is imperative. This involves a meticulous adherence to healthcare data regulations and privacy laws, with a particular emphasis on safeguarding patient information. Paramount among these regulations is the Health Insurance Portability and Accountability Act (HIPAA), which sets stringent standards for the protection and confidentiality of healthcare data. This dedication to regulatory compliance is a cornerstone in building trust within the healthcare ecosystem, instilling confidence among patients, healthcare providers, and stakeholders.

The seamless integration of 5G into healthcare necessitates a holistic approach, addressing infrastructure, security, interoperability, and compliance. Table 9.1 provides an overview of the current status of each technological prerequisite, along with the key technologies employed in them. By meeting these technological prerequisites, healthcare providers can harness the full potential of 5G technology to enhance patient care, streamline operations, and advance medical research. The technological prerequisites corresponding to each application in the integration of 5G in healthcare are illustrated in Figure 9.1.

**Table 9.1** Status and key technologies for 5G healthcare technological prerequisites [40, 41].

| Technological prerequisites | Implementation status | Key technologies used |
|---|---|---|
| Network infrastructure | Ongoing | 5G small cells, base stations, network slicing |
| Edge computing | Planning | Edge servers, fog computing, MQTT protocols |
| Security protocols | Completed | AES encryption, SSL/TLS protocols, blockchain |
| Interoperability standards | Ongoing | FHIR (fast healthcare interoperability resources), HL7, CCD (continuity of care document) |
| Device compatibility | Ongoing | IoMT devices, MQTT, CoAP (constrained application protocol) |
| Quality of service (QoS) management | Completed | Network slicing, quality of service algorithms |
| Data management and storage | Ongoing | Cloud infrastructure (AWS, Azure), hadoop, DICOM (digital imaging and communications in medicine) |
| Regulatory compliance | Ongoing | HIPAA compliance, GDPR (general data protection regulation) |

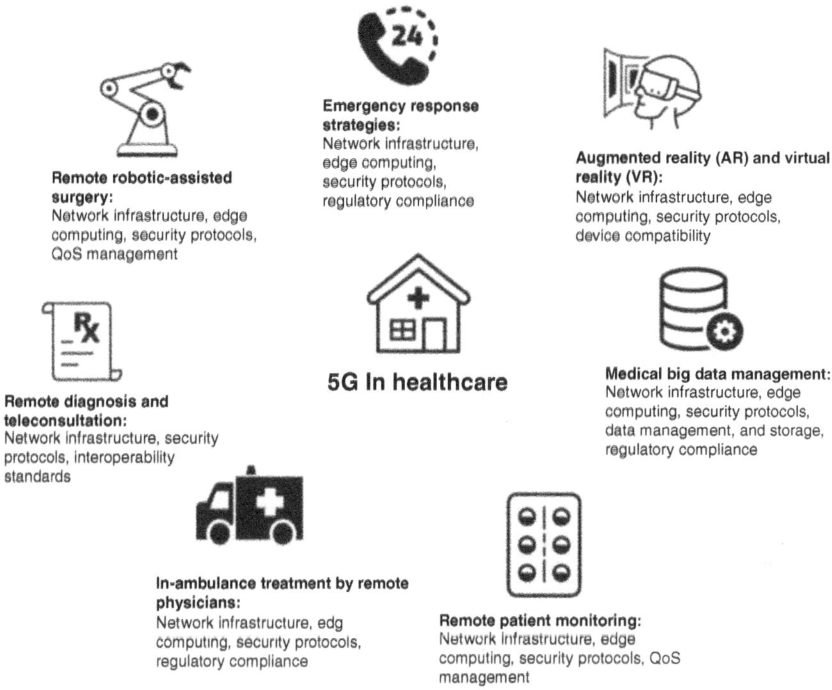


**Figure 9.1** The technological prerequisites corresponding to each application in the integration of 5G in healthcare.

## 9.4 Challenges in 5G Integration in Healthcare

The integration of 5G technology into healthcare systems holds immense promise for revolutionizing medical services. However, this technological evolution is not without its formidable challenges. The challenges in the integration of 5G in healthcare span various domains, from the substantial costs associated with infrastructure development to intricate concerns related to security, privacy, and regulatory compliance [10, 13, 22]. As the healthcare industry navigates this intricate landscape, understanding and addressing these challenges are imperative to unlock the full potential of 5G in transforming and elevating healthcare delivery.

- **Cost Implications:** The integration of 5G technology into healthcare infrastructure presents formidable challenges, notably the high initial setup costs. The deployment of 5G necessitates substantial investments in establishing infrastructure, including small cells and base stations. Additionally, the adaptation of existing healthcare systems and devices to align with 5G standards incurs additional expenses. Overcoming these cost-related challenges is crucial for ensuring the widespread implementation of 5G technology in healthcare settings.

- **Security and Privacy Concerns:** The advent of 5G introduces a higher number of connected devices in healthcare, amplifying security vulnerabilities. This increased connectivity raises concerns about the susceptibility of healthcare systems to cyber threats. Patient data security and privacy become paramount, requiring the implementation of robust encryption and authentication mechanisms. Addressing these concerns is imperative to instill confidence in patients and healthcare providers regarding the secure integration of 5G into healthcare environments.
- **Regulatory Challenges:** The integration of 5G in healthcare encounters complex regulatory challenges. Compliance with evolving regulations and standards pertaining to healthcare data privacy and telecommunications is essential. Navigating the intricate regulatory landscapes that vary across regions and countries adds an additional layer of complexity. Successful implementation requires a meticulous approach to align with regulatory requirements, ensuring seamless adherence to diverse standards governing healthcare and telecommunications.
- **Interoperability Issues:** Seamless integration and interoperability between diverse healthcare systems, devices, and platforms pose significant challenges in the 5G era [42]. Standardizing communication protocols becomes imperative to facilitate smooth data exchange among different components of the healthcare ecosystem. Ensuring that various elements within the healthcare infrastructure can communicate seamlessly is essential for realizing the full potential of 5G in improving healthcare services.
- **Network Reliability and Latency:** The consistent and reliable network connectivity that 5G promises is crucial for healthcare applications. However, ensuring such reliability, poses challenges, particularly in scenarios where latency is a critical factor. Applications such as remote surgery and real-time patient monitoring demand minimal latency to operate effectively. Addressing concerns related to network reliability and latency is essential to guarantee the success of critical healthcare interventions in a 5G-enabled environment.
- **Energy Consumption in WBAN Sensors:** Wireless Body Area Network (WBAN) healthcare systems introduce challenges related to energy consumption in lightweight sensors [43]. Sensors located on or within the body have limited energy, and replacing them can be inconvenient or painful for the patient. Furthermore, the increasing number of lightweight sensors may lead to data congestion on the radio network. Potential solutions include leveraging edge cloud and computation for local data processing, aiming to reduce latency while extending the lifespan of WBAN sensors.
- **Server-Side Challenges:** On the server side, challenges in 5G-enabled healthcare systems include coverage issues and effective data management.

The coverage of 5G may experience drops more frequently than 4G networks due to smaller cell tower coverage radii. Ensuring consistent long-range communication between patients, doctors, and medical servers is an ongoing concern. Effective data management involves coordinating data generated by various networks to guarantee the accessibility and completeness of patient information, enabling accurate diagnoses by healthcare professionals.

- **Spectrum Allocation:** Efficiently managing the limited radio frequency spectrum is a pivotal challenge in the integration of 5G into healthcare. Allocating spectrum resources to 5G networks and avoiding interference are critical aspects to maintain high-speed and reliable connectivity. The effective utilization of spectrum resources is essential for supporting the data-intensive requirements of healthcare applications enabled by 5G technology.
- **Dynamic Spectrum Sharing:** The concept of dynamic spectrum sharing presents challenges in optimizing the utilization of available frequencies in 5G networks. Developing technologies for dynamic spectrum sharing involves ensuring efficient allocation among various services and industries. Challenges arise in mitigating interference and achieving a harmonious spectrum-sharing environment. Addressing these challenges is crucial to enable seamless and effective utilization of available frequencies in the dynamic spectrum-sharing landscape of 5G.

Navigating these challenges and critical considerations demands a multidisciplinary approach involving collaboration between telecommunication experts, healthcare professionals, regulatory bodies, and technology developers. The successful integration of 5G into healthcare relies on addressing these challenges to unlock the full potential of advanced connectivity and technology-driven healthcare solutions. Table 9.2 outlines the challenges, current status, solutions, and metrics involved in the integration of 5G technology in healthcare.

## 9.5 Conclusion

In conclusion, the fusion of 5G technology and healthcare holds immense promise, offering a myriad of transformative applications that can significantly enhance the efficiency and effectiveness of medical services. The outlined applications, from remote surgeries to virtual consultations, underscore the potential to bridge geographical gaps and provide timely healthcare services. However, the successful implementation of 5G in healthcare necessitates careful consideration of various technological prerequisites, including robust network infrastructure, edge computing capabilities, stringent security protocols, and adherence to interoperability standards.

**Table 9.2** Challenges, status, solutions, and metrics in 5G healthcare integration [10, 13, 22].

| Challenges faced | Status | Proposed solutions | Monitoring and evaluation metrics |
|---|---|---|---|
| Cost implications | Ongoing; high initial setup costs | Explore public-private partnerships for funding<br>Research cost-effective infrastructure alternatives<br>Incentives for healthcare institutions adopting 5G | Return on investment (ROI)<br>Cost savings achieved through 5G integration<br>Percentage reduction in infrastructure setup costs |
| Security and privacy concerns | Ongoing; increased security risks | Implement robust encryption and authentication<br>Regular security audits and updates<br>Education and awareness programs for stakeholders | Number of security incidents or breaches<br>Percentage improvement in data security measures<br>User satisfaction with data privacy measures |
| Regulatory challenges | Ongoing; complex regulatory landscape | Collaborate with regulatory bodies for compliance<br>Establish standardization in data privacy regulations<br>Continuous monitoring of regulatory changes | Regulatory compliance score<br>Time taken to adapt to new regulations<br>Number of successful regulatory audits |
| Interoperability issues | Ongoing; hinders seamless integration | Develop and adopt standardized communication protocols<br>Encourage industry-wide collaboration for standards<br>Invest in interoperability testing and certification | Interoperability success rate<br>Average time for data exchange between different platforms<br>User feedback on system interoperability |

| | | | |
|---|---|---|---|
| Network reliability and latency | Ongoing; critical for healthcare apps | Prioritize low-latency solutions for critical apps | Network uptime and reliability |
| | | Optimize network architecture for reliability | Average latency for critical healthcare applications |
| | | Implement redundancy for mission-critical applications | Percentage reduction in network downtime |
| Energy consumption in WBAN sensors | Ongoing; limited sensor energy | Local data processing through edge cloud | Energy consumption per data transmission |
| | | Research and adopt energy-efficient sensor solutions | Lifespan extension of WBAN sensors with new solutions |
| | | Develop protocols for optimizing sensor data transfer | User satisfaction with sensor energy efficiency |
| Server-side challenges | Ongoing; coverage and data management | Improve 5G tower coverage for consistent connectivity | Coverage gaps and drops in 5G network |
| | | Invest in advanced data management and analytics | Data accessibility and completeness metrics |
| | | Employ AI for efficient data processing and retrieval | User satisfaction with server-side improvements |
| Spectrum allocation | Ongoing; critical for high-speed connectivity | Collaborate with regulatory bodies for spectrum allocation | Efficient use of allocated spectrum resources |
| | | Optimize spectrum use through advanced technologies | Minimization of interference issues |
| | | Develop dynamic spectrum allocation strategies | User satisfaction with high-speed connectivity |
| Dynamic spectrum sharing | Ongoing; challenges in optimization | Develop advanced spectrum-sharing technologies | Spectrum-sharing efficiency |
| | | Implement interference mitigation techniques | Mitigation of interference issues during spectrum sharing |
| | | Promote industry collaboration for harmonious sharing | User satisfaction with dynamic spectrum-sharing solutions |

Despite these technological strides, challenges persist. The financial implications of adopting 5G, coupled with concerns related to security, privacy, and regulatory compliance, pose challenges that require strategic solutions. Furthermore, the energy consumption associated with WBAN sensors demands innovative approaches to mitigate environmental impact. As the healthcare industry endeavors to overcome these challenges, collaboration between technology developers, healthcare providers, policymakers, and regulatory bodies becomes paramount. By fostering a collaborative ecosystem, stakeholders can work together to address the obstacles in the integration of 5G in healthcare, paving the way for a future where advanced connectivity contributes to improved patient outcomes, streamlined medical processes, and a more resilient healthcare infrastructure.

## References

1 Jagarapu, J. and Savani, R.C. (2021). A brief history of telemedicine and the evolution of teleneonatology. In *Seminars in Perinatology* Elsevier. 45 (5): 151416. WB Saunders.
2 Hjelm, N. and Julius, H.W. (2005). Centenary of tele-electrocardiography and telephonocardiography. *Journal of Telemedicine and Telecare* 11 (7): 336–338.
3 Ting, D.S. et al. (2020). Artificial intelligence, the internet of things, and virtual clinics: ophthalmology at the digital translation forefront. *The Lancet Digital Health* 2 (1): e8–e9.
4 Hong, Z. et al. (2020). Telemedicine during the COVID-19 pandemic: experiences from Western China. *Journal of Medical Internet Research* 22 (5): e19577.
5 Zhou, B. et al. (2020). Construction of 5G all-wireless network and information system for cabin hospitals. *Journal of the American Medical Informatics Association* 27 (6): 934–938.
6 Hollander, J.E. and Carr, B.G. (2020). Virtually perfect? Telemedicine for COVID-19. *New England Journal of Medicine* 382 (18): 1679–1681.
7 Wickham, L. et al. (2020). The impact of COVID policies on acute ophthalmology services—experiences from Moorfields Eye Hospital NHS Foundation Trust. *Eye* 34 (7): 1189–1192.
8 Bourdon, H. et al. (2020). Teleconsultation in primary ophthalmic emergencies during the COVID-19 lockdown in Paris: experience with 500 patients in March and April 2020. *Journal Français d'Ophtalmologie* 43 (7): 577–585.
9 Nurses, I.C.O. (2021). ICN says 115,000 healthcare worker deaths from COVID-19 exposes collective failure of leaders to protect global workforce. ICN Geneva, Switzerland.

**10** Moglia, A. et al. (2022). 5G in healthcare: from COVID-19 to future challenges. *IEEE Journal of Biomedical and Health Informatics* 26 (8): 4187–4196.

**11** Li, J.-P.O. et al. (2021). Digital technology, tele-medicine and artificial intelligence in ophthalmology: a global perspective. *Progress in Retinal and Eye Research* 82: 100900.

**12** Ullah, H. et al. (2019). 5G communication: an overview of vehicle-to-everything, drones, and healthcare use-cases. *IEEE Access* 7: 37251–37268.

**13** Qureshi, H.N. et al. (2020). Service level agreements for 5G and beyond: overview, challenges and enablers of 5G-healthcare systems. *IEEE Access* 9: 1044–1061.

**14** Gupta, P. and Ghosh, M. (2019). Revolutionizing healthcare with 5G. *Telecom Business Review: SITM Journal* 12: 41.

**15** Li, D. (2019). 5G and intelligence medicine—how the next generation of wireless technology will reconstruct healthcare? *Precision Clinical Medicine* 2 (4): 205–208.

**16** US Food and Drug Administration. (2013). Radio frequency wireless technology in medical devices-guidance for industry and food and drug administration staff.

**17** Ye, J. (2020). The role of health technology and informatics in a global public health emergency: practices and implications from the COVID-19 pandemic. *JMIR Medical Informatics* 8 (7): e19866.

**18** Zheng, J. et al. (2020). 5G ultra-remote robot-assisted laparoscopic surgery in China. *Surgical Endoscopy* 34: 5172–5180.

**19** Chen, H. et al. (2021). Application of 5G technology to conduct real-time teleretinal laser photocoagulation for the treatment of diabetic retinopathy. *JAMA Ophthalmology* 139 (9): 975–982.

**20** Gupta, R. et al. (2019). Tactile-internet-based telesurgery system for healthcare 4.0: an architecture, research challenges, and future directions. *IEEE Network* 33 (6): 22–29.

**21** Meshram, D.A. and Patil, D.D. (2020). 5G enabled tactile internet for tele-robotic surgery. *Procedia Computer Science* 171: 2618–2625.

**22** Zhang, D. et al. (2019). 5G-enabled health systems: solutions, challenges and future research trends. In: *2019 ITU Kaleidoscope: ICT for Health: Networks, Standards and Innovation (ITU K)*. IEEE.

**23** Wang, J. et al. (2021). Application of a robotic tele-echography system for COVID-19 pneumonia. *Journal of Ultrasound in Medicine* 40 (2): 385–390.

**24** Duan, S. et al. (2021). A 5G-powered robot-assisted teleultrasound diagnostic system in an intensive care unit. *Critical Care* 25 (1): 1–9.

**25** Zhang, Y. et al. (2021). Ensemble learning based postpartum hemorrhage diagnosis for 5G remote healthcare. *IEEE Access* 9: 18538–18548.

**26** Usman, M.A., Philip, N.Y., and Politis, C. (2019). 5G enabled mobile healthcare for ambulances. In: *2019 IEEE Globecom Workshops (GC Wkshps)*. IEEE.

**27** Zhai, Y. et al. (2021). 5G-network-enabled smart ambulance: architecture, application, and evaluation. *IEEE Network* 35 (1): 190–196.

**28** Peralta-Ochoa, A.M. et al. (2023). Smart healthcare applications over 5G networks: a systematic review. *Applied Sciences* 13 (3): 1469.

**29** Sivasangari, A. et al. (2021). Big data analytics for 5G-enabled IoT healthcare. In: *Blockchain for 5G-Enabled IoT: The New Wave for Industrial Automation*, 261–275. Springer.

**30** Gia, T.N. et al. (2017). Low-cost fog-assisted health-care IoT system with energy-efficient sensor nodes. *2017 13th International Wireless Communications and Mobile Computing Conference (IWCMC)*. IEEE, US.

**31** He, S. et al. (2017). Proactive personalized services through fog-cloud computing in large-scale IoT-based healthcare application. *China Communications* 14 (11): 1–16.

**32** Ali, S. and Ghazal, M. (2017). Real-time heart attack mobile detection service (RHAMDS): an IoT use case for software defined networks. *2017 IEEE 30th Canadian Conference on Electrical and Computer Engineering (CCECE)*. IEEE, Windsor, Ontario, Canada.

**33** Constant, N. et al. (2017). Fog-assisted wiot: a smart fog gateway for end-to-end analytics in wearable internet of things.

**34** Li, L., Ota, K., and Dong, M. (2018). Deep learning for smart industry: efficient manufacture inspection system with fog computing. *IEEE Transactions on Industrial Informatics* 14 (10): 4665–4673.

**35** Mahmud, R., Koch, F.L., and Buyya, R. (2018). Cloud-fog interoperability in IoT-enabled healthcare solutions. *Proceedings of the 19th International Conference on Distributed Computing and Networking*. Varanasi, India.

**36** Barik, R.K. et al. (2018). FogLearn: leveraging fog-based machine learning for smart system big data analytics. *International Journal of Fog Computing (IJFC)* 1 (1): 15–34.

**37** Rajkomar, A. et al. (2018). Scalable and accurate deep learning with electronic health records. *npj Digital Medicine* 1 (1): 18.

**38** Sahoo, P.K. et al. (2018). SLA based healthcare big data analysis and computing in cloud network. *Journal of Parallel and Distributed Computing* 119: 121–135.

**39** Reddy, S. (2018). Use of artificial intelligence in healthcare delivery. In: *eHealth-Making Health Care Smarter*. IntechOpen.

**40** Georgiou, K.E. et al. (2021). 5G use in healthcare: the future is present. *JSLS: Journal of the Society of Laparoscopic & Robotic Surgeons* 25 (4): 1–22.

**41** Anwar, S. and Prasad, R.J.W.P.C. (2018). Framework for future telemedicine planning and infrastructure using 5G technology. *Wireless Personal Communications* 100: 193–208.

**42** Akpakwu, G.A. et al. (2017). A survey on 5G networks for the internet of things: communication technologies and challenges. *IEEE Access* 6: 3619–3647.

**43** Jones, R.W. and Katzis, K. (2018). 5G and wireless body area networks. *2018 IEEE Wireless Communications and Networking Conference Workshops (WCNCW)*. IEEE, Barcelona.

# 10

# Dynamic Resource Orchestration for Computing, Data, and IoT in Networked Systems: A Data-Centric Approach

*Suresh Limkar*[1], *Mohammad Alamgir Hossain*[2], *Sherif Tawfik Amin*[2], and *Yasir Ahmad*[3]

[1] *Department of Artificial Intelligence & Data Science, AISSMS Institute of Information Technology, Pune, Maharashtra, India*
[2] *Department of Computer Science, College of Computer Science & information Technology, Jazan University, Jazan, Kingdom of Saudi Arabia*
[3] *Department of Information Technology & Security, College of Computer Science & Information Technology, Jazan University, Jazan, Kingdom of Saudi Arabia*

## 10.1 Introduction

The careful coordination of resources is crucial nowadays in linked systems to ensure best methods and quick responses in computing, data administration, and the Internet of Things (IoT). One of the creative ideas that developed to solve this important problem is focusing on data. The idea of this strategy is that data should be at the center of organizing resources, and guiding flexible decisions and allocations based on changing needs and outside factors. Basically, dynamic resource coordination (DRC) includes the flexible use of computer resources, effective ways to handle data, and smoothly adding IoT devices to linked systems [1, 2]. How modern programs work is making the old, stuck concepts of giving out resources not enough. So, to use today's technology fully, shifting to a data-focused way becomes really important.

The strategic arrangement and flow of data [3] is given top priority in the data-centric approach since it serves as the crucial foundation for resource orchestration. This methodology sees data as a vibrant and critical component that helps systems adapt quickly to shifting demands and workloads. This is particularly significant in the field of computers, as different applications require different amounts of processing power and access to data [4].

*Current and Future Cellular Systems: Technologies, Applications, and Challenges*, First Edition.
Edited by Garima Chopra, Suhaib Ahmed, and Shalli Rani.

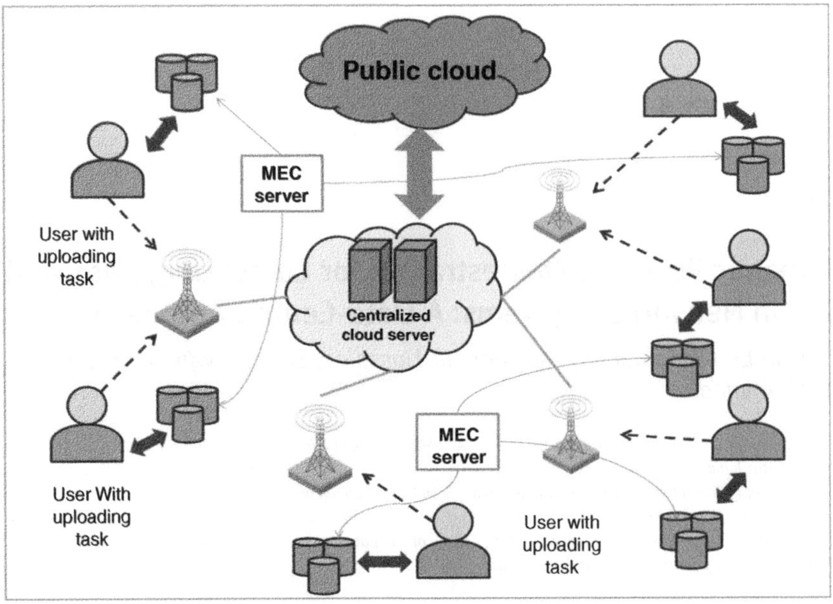


**Figure 10.1** Effective architecture for resource allocation in networks.

Figure 10.1 shows how things are set up across a system. It likely includes parts to increase how much is moving through, lessen the times waiting, and make the whole thing work best. The drawing may show how different pieces interact together using rules and ways to share what's needed in a streamlined way. The goal is to keep the system running without problems so it can do as much as possible.

The picture in Figure 10.1 shows how different sections connect through equations and repeating circles to enhance how resources are shared flexibly. Working together this way is important for doing better overall, lowering times to wait, and handling how busy it is in a good way. This figure stresses how important equations and computers are for making choices quickly, as they said in ref. [5]. That source talks about how important it is for things to work well to run the network smoothly.

These complex equations aid in evenly sharing computer jobs and improving how information moves around by using information from information. This helps make using technology better, saves money, and protects what we have. Figuring things out this way boosts how linked things work together and strengthens how happy people are when they use it and how well it does what it's meant to.

### 10.1.1 Motivation

This investigation explores the complex relationships between computing, data, and the IoT in connected systems. Though these areas differ, they share increasing commonalities. This demands a unified approach when coordinating assets. Emerging technologies such as internet-connected appliances, cloud computing, edge and fog computing, and data-focused applications present prospects and barriers. Our goal is to disentangle intricacies, find shared subjects, and apply data as a basic principle for adaptable asset coordination.

### 10.1.2 Objectives

- **Foundational Understanding:** Gain a thorough grasp of the fundamental ideas behind dynamic resource orchestration (DRO), highlighting its importance in the context of modern networked systems.
- **Computing Across Connected Systems:** Investigate different approaches to computing such as edge, cloud, and fog. Explore how to thoughtfully integrate these computing assets for optimal outcomes.
- Data is vital for coordinating resources. Discuss how analyzing information leads to better handling of details, overseeing things, and sharing assets. When using an information-focused approach, make sure to cover the important privacy and security issues.

## 10.2 Dynamic Resource Orchestration: Foundations

### 10.2.1 Resource Orchestration Concepts

Policy-driven orchestration offers a layer of control and predictability to resource management, confirming that resource allocations align with organizational priorities and legal standards.

When dealing with large amounts of data, keeping data close to the resources using it is important. This idea shows how latency needs to be low and access times optimized by placing data near computing resources that require it [6]. Solutions for effective data localization overall improve the performance of systems, especially those working with huge datasets like machine learning (ML) and data analysis [7]. Also, the idea of self-fixing systems increases how strong connected systems are. In dynamic environments, errors or problems are unavoidable. Self-healing technologies allow faults to be found and fixed automatically so functioning continues without interruption from people [8]. Together, these methods strengthen and make resource coordination systems more reliable.

### 10.2.2 Dynamic Resource Orchestration's Evolution

DRC has progressed throughout history due to revolutionary changes brought about by the IoT, data management, and computing environments. As the picture shows. Resource coordination has developed greatly over the years because of the growing complexity of networked systems, changing needs, and technological advances. Earlier, stationary allocation models where computing resources were assigned according to expected workloads were the main focus of resource coordination [9].

Resource management techniques were improved even more through policy-directed coordination. Guidelines offer oversight to direct choices about how resources are shared. These guidelines were based on what the organization primary objectives, the need to follow regulations, and the special requirements of different applications. With the introduction of control and customization [13] from this progress, resource coordination could better match both rules and larger business goals. At the same time, resource coordination became much more complex because connected things were everywhere. The evolution now had to think about what different connected things could do and need to connect. To make sure computer networks could change to match changes in data from smart sensors and how they connect, DRC came up to easily include these devices [14, 15] (Figure 10.2).

Data is used more and more to make choices, so methods that focus on data grew as ways to share resources. Putting data near computing power to boost how well things work, especially for tasks that use a lot of data like ML, became really important. How resources are shared will likely keep improving as new things such as 5G wireless networks and edge computing continue changing computing [19, 20]. Resource sharing will keep evolving because of the need for more exact control, fast responses, and good use of what's available. This ensures computer networks can adapt to problems, adjust to demands, and meet what new tech requires.

### 10.2.3 Importance of a Data-Centric Perspective

It is very important to use a data-focused plan when managing changing technology needs flexibly. Older models often based choices on preset rules, which did not work well when needs changed. With a data-focused plan, the information itself becomes the main thing to consider. Systems can then make good judgments based on real-time facts when they see data as the key part of managing technology resources [16]. This recognizes that different apps need varying amounts of computing power and access information in different ways. As a result, managing

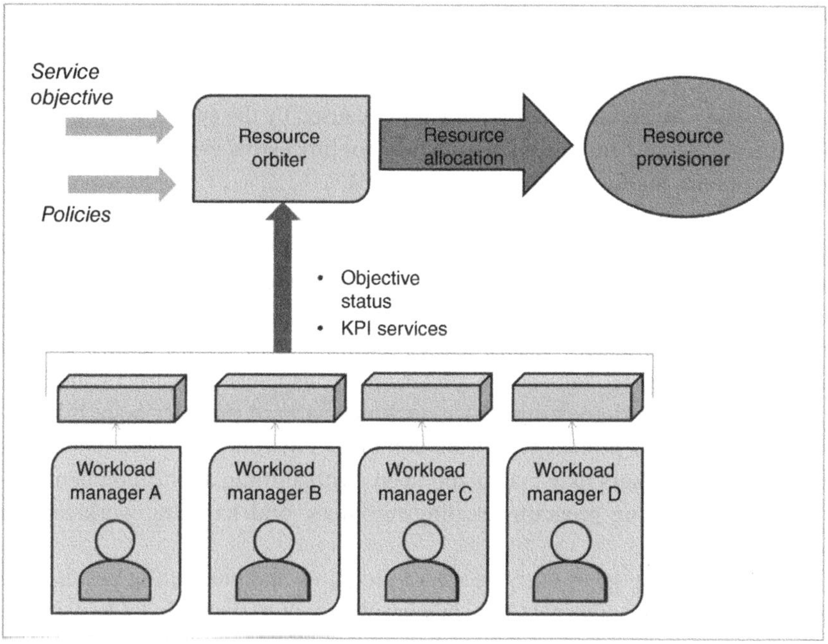


**Figure 10.2** Representation of dynamic resource allocation.

resources becomes more adjustable, matching technology to changing app needs. Furthermore, a data-focused plan ensures resources are used well to handle and examine the large amount of information made by many connected things, where devices constantly make data [17]. The plan uses smart algorithms [18] that think about what data is urgent and relevant and needs processing to efficiently share out resources.

The move from rigid designs to more reactive and adjustable approaches forms the root of lively computer coordination. It is impossible to exaggerate how significant a data-directed plan is because it establishes sensible management of means. Accepting these basics is pivotal for engineering computer frameworks that are productive, expandable, and strong enough to meet the necessities of the computerized time, particularly as associated frameworks keep on developing.

## 10.3 Computing in Networked Systems

### 10.3.1 Cloud Computing Paradigm

Within the area of interconnected systems computing, cloud computing displays a shifted approach altering how companies allocate, utilize, and control their IT assets. Essentially, cloud computing is an online model offering computing services when requested. These services can involve everything from storage and handling ability to apps and analytics. Because it can provide scalable, flexible, and cost-effective solutions to the difficulties caused by the ever-changing nature of modern programs and the increasing need for processing assets, the cloud computing model is highly significant (Figure 10.3).

Sharing computing resources across many users through resource pooling is a key feature of cloud computing. As a result, businesses can access and use resources in a pay-as-you-go manner, leading to economies of scale. Cloud computing is inherently flexible, allowing users to adjust resource amounts in response to demand. This allows for optimal resource use and less need to over-provide [17]. Self-service provisioning also lets consumers independently provide and oversee computer resources without provider help. It boosts responsiveness and proficiency in the computer environment while increasing adaptability too. Users may match app needs with shifting application demands by rapidly launching programs, scaling resources, and handling workloads in real time.

Cloud computing gives users access to programs and computing power from anywhere using the internet. This lets groups work globally and adapt fast to

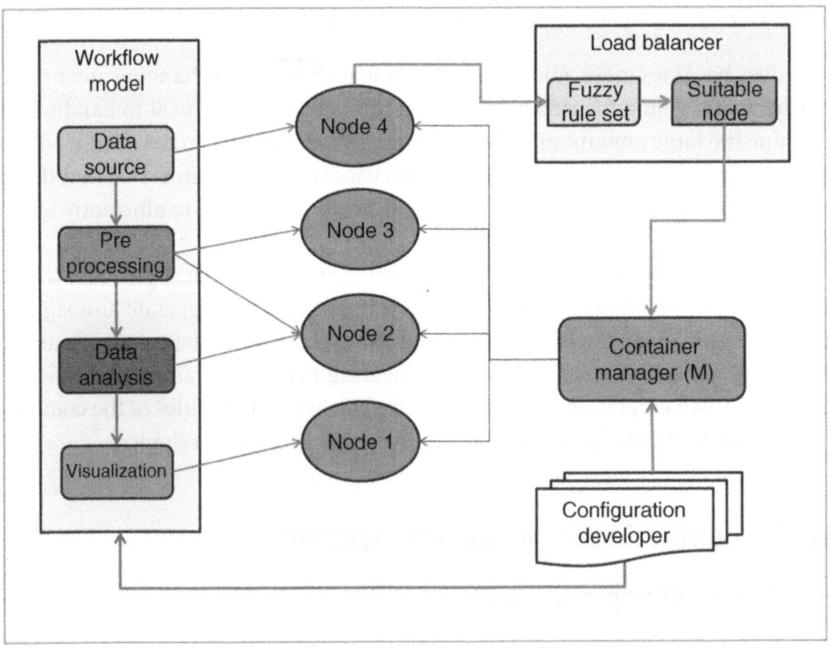


**Figure 10.3** Representation of allocating dynamic resources in container-based services.

changing needs. It helps with sharing, flexibility, and remote access to computer power. The cloud makes computing available no matter where you are. It breaks down place limits and encourages networking computers more. Cloud computing also introduces different service models. Software as a Service or SaaS, shares apps online for a fee. Platform as a Service (PaaS) lets makers build, put out, and run apps without worrying about the bottom parts. Infrastructure as a Service (IaaS) offers computing parts over the internet as well. With many service models to pick from, companies can choose the right level for their needs. This makes using resources fit what each group wants.

Additionally, the computing paradigm supported by the cloud [21, 22] is essential to developing fresh technologies such as ML, big data investigation, and Web of Things (IoT). Cloud means the perfect setting for processing and analyzing huge datasets, teaching ML styles, and fulfilling the many needs of IoT deployments because of their ability to scale and adaptability. The cloud computing paradigm signifies a major change in how computing means are allocated, accessed, and controlled within networks. Its impact extends beyond new technology [21, 22] to incorporate how businesses organize their IT functions and strategies. Cloud computing, with its focus on scalability, self-provisioning, universal access, and also on a variety of service versions, is a key part of developing computing within networked systems, offering a basis for resource performance, creativity, and mobility.

### 10.3.2 Edge Computing and Fog Computing

The world of computing on connected systems progresses rapidly, with edge computing and fog computing emerging as fresh ideas. By locating processing power nearer information sources and users, these strategies beat the downsides of standard cloud computing models. They also provide instant, real-time handling, and overall framework proficiency. Edge computing is characterized by decentralizing computing assets closer to where information is made and gadgets are utilized. Not at all like conventional cloud computing, has which included sending information to a central cloud server for preparing, edge computing moves handling closer to system boundaries, typically near information origins (Figure 10.4).

Edge and fog computing are developing as ways to spread processing power across more places. This shifts computing to models that are more adjustable and useful for linked systems. These approaches [23–26] lower how much bandwidth is needed, make security better for the whole system, and meet the needs of programs that must react quickly. They do this by handling data where it starts and near the devices people use. Edge and fog computing will strongly affect how computing systems are composed going forward. This is because more and more "things" will connect to the internet and the demand for instant processing will

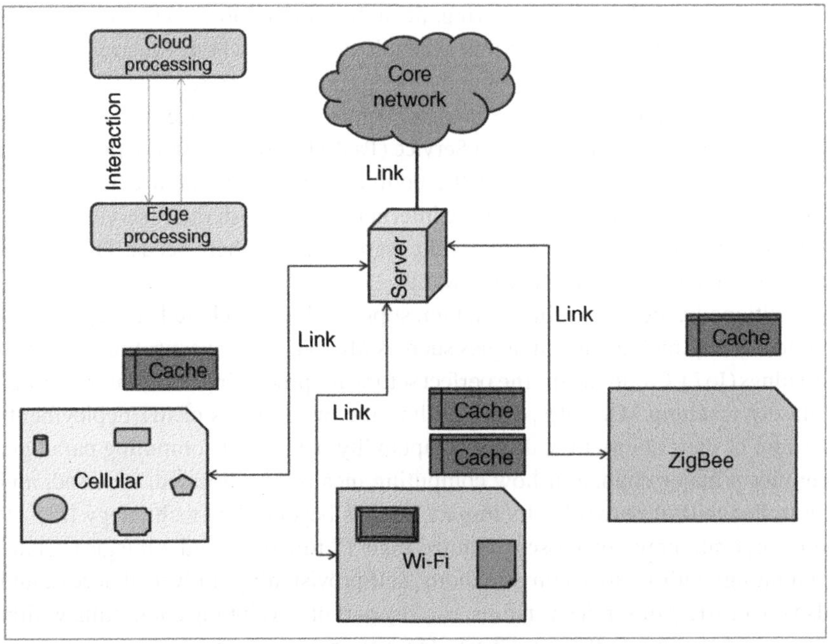


**Figure 10.4** Mobile edge cloud computing and big data analytics.

keep rising. They provide important benefits as a supplement to the standard cloud computing approach.

### 10.3.3 Integration of Computing Resources

Different types of computer systems must work together smoothly to meet modern needs. An important part of this is integrating resources across networks. This integration promotes sharing, using what each has to offer efficiently, and having setups that can change as wanted. A complex but important method called integration unites different computer ideas, hardware, and software to make one system that works well (Figure 10.5).

This [27] encourages portability and speeds up how applications can change in size, creating a more responsive and adaptable computing environment. To further improve resource use, container directing solutions especially automate deploying, scaling, and taking care of containerized applications.

Algorithms powered by AI and ML [10–12] learn from data to estimate workloads, assign resources in the best way, and make choices automatically. These technologies help manage resources intelligently so environments run at their peak and change smoothly. Any integrated computing setup must consider

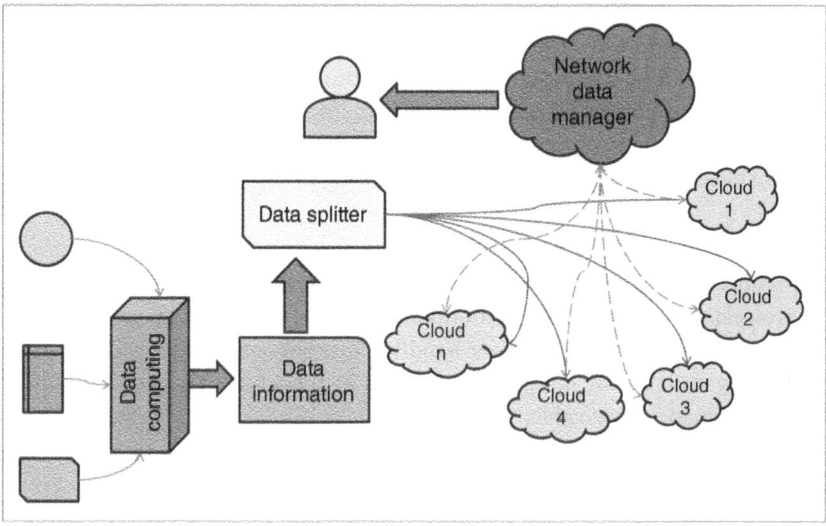


**Figure 10.5** Data computing and resource manager.

security and rules. It must smoothly combine security features such as threat spotting, access control, and encryption. Governance models and compliance frameworks also make sure the integration follows legal rules, protects sensitive data, and keeps user and stakeholder trust.

## 10.4 Data-Centric Orchestration

Data insights guide resource choices in "Data-Centric Orchestration." This strategic method focuses on data and ensures computing resources flexibly adjust to apps as demands vary. This happens in systems linking computing, data handling, and internet-linked items. Insights from live data are key. Three main parts manage this process. Data guides how resources share out and shift. It also steers how processing and care of data happens. Keeping data safe and private is the third concern. Together, these make sure the whole network serves users well.

### 10.4.1 Data-Driven Resource Allocation

#### 10.4.1.1 Data-Driven Decision-Making

At times the information helped decide where to send things needed. The setup changed how it used its thinking parts based on the jobs, what was around, and what the data was like at that moment. By sending its processing capabilities only where and when needed, this approach proved more effective.

#### 10.4.1.2 Dynamic Scaling

The way data helps distribute tools has flexible adjusting as one of its big feature. To make sure things use tools in the best way, things can adjust tools up or down easily to the changing work needs. This is very important because what tools are needed might change a lot over time, especially with cloud finding where what tools are needed might change by a lot over time.

#### 10.4.1.3 Perceptive Formulas

Smart programs that review new information and make clever choices are the base for information-guided asset distribution. These programs think about things like the sort of information, the sum of handling intensity expected, and past usage patterns. By utilizing ML designs to anticipate workloads later on, proactive asset adjusting is made conceivable.

#### 10.4.1.4 Customization and Adaptability

Some programs and websites require various levels of resources. Data-driven resource sharing allows customization to ensure certain needs are met. The system adapts to different workloads, providing a personalized way of allocating resources based on the traits of the information being used.

### 10.4.2 Data Processing and Management

#### 10.4.2.1 Data Locality and Optimization

Effective data processing and management form the basis of data-focused coordination. Keeping data near the computing power that requires it is important, highlighted by the idea of data placement. This cuts down wait time, quickens how long it takes to get data, and ultimately boosts how well the system runs.

#### 10.4.2.2 Techniques for Data Movement

The strategy considers how data is spread and moved. Data could be shared between network parts, saved briefly, or copied, depending on what is needed. These plans make sure data can be gotten to quickly where and when it is required.

#### 10.4.2.3 Data Lifecycle Management

Storing and deleting information at each step from when it is made and changed to when it is kept or thrown away is important for working with information the right way. Every part of dealing with information, like how long to keep it, putting it away after using it, and getting rid of it, follows rules and ways of doing things. This helps use space best and makes sure to follow what the law says.

#### 10.4.2.4 AI and Data Analytics Integration

Data-driven systems commonly employ complicated evaluations and man-made brainpower programs in information handling. These programs leverage the wealth of accessible information to robotize choice-creating, foresee potential results, and uncover important understandings [28, 29]. Joining synthetic knowledge and information examination can significantly amplify the benefits extracted from the facts on hand.

### 10.4.3 Security and Privacy Considerations

#### 10.4.3.1 Completely Encryption

Data-centric orchestration places a high priority on security. Data is kept private and secure all the way from the place of origin to the destination thanks to end-to-end encryption. In the process of transmitting data, this secures confidential information and guards against unauthorized access.

#### 10.4.3.2 Identity and Access Management

Access to modify data is carefully managed. Strict rules define who can see or change which information. The system uses identification, permissions, and logins tailored for each person's job. This ensures the right people are able to do their work.

#### 10.4.3.3 Safe Data Processing

Working with information requires keeping it safe. Strategies can help make sure data stays secure even when being used. Special protections may surround the data or split it between parties for joint work. This is crucial for sensitive details that must stay shielded while being studied.

#### 10.4.3.4 Regulatory Standard Compliance

At times, how data-driven coordination works relies upon the business and area. It must stick to different administrative standards and constant necessities. This expects adhering to protection laws, industry-explicit rules, and information assurance standards. The coordination system incorporates consistence forms to lessen lawful and administrative stresses.

#### 10.4.3.5 Privacy-Preserving Techniques

Some techniques could help keep information private when it is analyzed. Homomorphism encryption and differential privacy allow data to be studied without sharing secret details. These methods make sure a fair middle ground exists between how useful the data is and how much an individual wants to keep private.

#### 10.4.3.6 Audit Trails and Monitoring

Different types of checks and watches are set up to keep people responsible and open. Tools see how the system acts overall and what gets changed or used from the information. Step logs help with studying prints and looking into things if bad things happen to security or breaks happen. Understanding data orchestration is a plan that changes how computer resources are handled by showing how important data is when choices are made. It combines using data well and keeping it organized, changing how resources are shared, and having good protection of information and privacy to make computing safe, adaptable, and reactive when systems work together over a network.

## 10.5 IoT Integration

The IoT is connecting more devices than ever before. This new kind of network makes machines able to share information and work together in new ways. However, integrating so many devices brings challenges too. We must think carefully about how to use resources efficiently across different types of connected things. Each has its own design and problems to solve. Still, if we address these issues, networks of devices can help computing and data handling in new areas.

### 10.5.1 Overview of IoT Architecture

Some devices, sensors, and machines can connect with computers and each other in small ways. This new way linked all kinds together into a big system. Most use three levels: the edge level where the devices are, the middle level that moves information between edges and clouds, and the cloud level where large computers help make sense of everything.

- **The Outermost Edge:** Many internet-connected machines, sensors, and automatic mechanisms inhabit this physical realm. Gadgets of all sorts collect facts about their surroundings, do some preliminary figuring out if needed, and then beam especially important information to the next level up.
- The network layer facilitates communication between edge devices and larger computing systems. It organizes how data moves between the multitudes of connected things. Routers and gateways within this strata guide information transfers and connections for the diverse array of internet-linked objects.
- The massive amounts of information generated by IoT gadgets are kept, inspected, and prepared additionally by the cloud layer. The capacity to develop and extend and the registering force expected to break down and store enormous measures of information are made conceivable by cloud registering assets.

## 10.5.2 IoT Resource Orchestration Challenges

### 10.5.2.1 Device Heterogeneity

Some devices need more energy than others to operate. They use different ways to connect and have varying abilities to process data. To organize how resources are shared, we must handle these differences. The many types of objects linked to the internet, from simple sensors to powerful computers, challenge how we coordinate what each needs. Effective management of this diversity is crucial for planning access to processing, networking, and power.

### 10.5.2.2 Scalability and Data Volume

The large amounts of data made by internet-connected things cause a big problem. Special systems that share and look at huge groups of facts and numbers quickly as things happen are needed. This is hard when there are lots of devices. By spreading the work to where the facts start and not making the main cloud computers do too much, edge computing must help with this issue.

### 10.5.2.3 Low-Latency and Real-Time Processing

IoT devices are often used in critical situations such as self-driving vehicles or automated factories. Sending data to faraway servers could cause delays in responding. Edge computing must help coordinate things for fast reactions, especially in emergencies. These devices often handle sensitive information from medical or home environments, so safeguarding devices, data, and networks against potential threats is crucial for effective resource sharing. Strong security measures, encryption, and usage policies are essential

### 10.5.2.4 Compatibility and Standards

Some gadgets work together well, others do not. Big companies make devices but do not always share how they talk to each other. This can cause issues when devices from different places try to work as a team. To fix this, experts suggest everyone agree to use the same language. If all our smart things speak the same way, they can share resources and help us better. Making common rules for all helps gadgets combine their powers.

## 10.5.3 Combining Data and Computing

Edge computing helps solve problems by using IoT resources. It makes the system work better by decreasing wait time, saving data traffic on networks, and handling information near where it starts at the edge. Edge computing working with IoT lets resources be used in better ways. This is important when information needs fast processing, like in situations where every second counts.

#### 10.5.3.1 Data-Centric Orchestration

Some concepts behind this method work well when including internet-connected devices. Methods that organize computer resources can be distributed flexibly depending on up-to-date facts from devices on the IoT. By focusing processing power where needed most right now, helps confirm the IoT system works most effectively and reactively.

#### 10.5.3.2 IoT with Machine Learning and AI

The coordination of tools is enhanced when things online are joined with machines studying and fake brains. Machines studying models can do the best allocation of assets, foresee how gadgets will act, and break down examples in information. For example, modern industrial IoT prevents equipment damage by utilizing machines and figuring out calculations to discover faults in machines and gear, extend maintenance schedules, and reduce downtime.

#### 10.5.3.3 Dynamic Resource Allocation

Sometimes resources need adjusting based on what devices are doing online. It helps if things linked to the IoT can get potential resources depending on how busy or connected they are at the time. Resources should change to suit what the IoT world needs right then, so things don't cost too much and everything works well.

Let's denote:

- $R(t)$: Resource demand at time $t$.
- $P(t)$: Predicted resource demand at time t from the ML model.
- $\theta$: Threshold or policy for decision making.

**Prediction:**

$$P(t) = \text{Predict}(R(t-1), R(t-2), \ldots, R(t-n))$$

**Decision Making:**

$$\text{Decision}(P(t), \theta) = \{\text{Increase Resources if } P(t) > \theta$$

$$\text{Maintain Resources if } P(t) \leq \theta\}$$

**Resource Allocation:**

Allocate_Resources(Decision)

**Prediction:**

$$P(t) = \text{Predict}(R(t-1), R(t-2), R(t-3))$$

**Decision Making:**

$$\text{Decision}(P(t), \theta) = \{\text{Increase Resources if } P(t) > \theta\}$$

#### 10.5.3.4 IoT Security Measures

Different parts of making things on the IoT talk to each other safely and deal with rules for keeping computers and information safe in general. How devices and information on the IoT are kept private uses proving who is using what and allowing what, encoding information so only the right people can see it, and following rules set by people in charge. Bringing all the parts together into a tidy, safe connected system combines these safeguards.

**HMAC Pseudo-Algorithm:**

**Inputs:**

- K: Secret key
- M: Message to be authenticated
- H: Hash function (e.g. SHA-256)

**Outputs:**

$$\text{HMAC}(K, M) : \text{HMAC value}$$

- **Key Padding:**
  - If the length of K is greater than the block size of the hash function, K is hashed to match the block size.
  - If the length of K is less than the block size, K is padded with zeros to match the block size.
- **Key XOR with Outer Pad:**
  - Create outer_pad by repeating the byte 0×5c for each byte in a block.
  - XOR each byte of K with the corresponding byte of outer pad.
- **Message Padding:**
  - If the length of M is greater than the block size, M is hashed to match the block size.
  - If the length of M is less than the block size, M is padded with zeros to match the block size.
- **Key XOR with Inner Pad:**
  - Create inner_pad by repeating the byte 0×36 for each byte in a block.
  - XOR each byte of K with the corresponding byte of inner_pad.
- **Concatenate and Hash:**
  - Concatenate the result from step 4 with the result from step 3.
  - Hash the concatenated result using the hash function H.
- **Concatenate and Hash Again:**
  - Concatenate the result from step 2 with the result from step 5.
  - Hash the concatenated result using the hash function H.
- **Output:**
  - The final hash value is the HMAC value.

Resource orchestration faces both possibilities and problems from the incorporation of IoT within networked systems. IoT's distinct architecture, with its layers of network, cloud, and edge, necessitates creative solutions to deal with heterogeneous devices, scalability problems, and real-time processing demands. An IoT ecosystem that is more responsive, effective, and safe is made possible by the combination of cutting-edge technologies like edge computing and ML with synergies with computing and data-centric orchestration.

## 10.6 Methodologies for Dynamic Resource Orchestration

Coordinating how data, tasks, and internet-connected devices share resources in connected systems requires adjusting how things are arranged to meet changing needs and make the most of what's available. There are three main ways to handle flexible resource management. ML uses examples to recognize patterns and predict resource demands. Optimization methods arrange resources in the best way using math. Hybrid models combine these approaches.

### 10.6.1 Methods of Machine Learning

#### 10.6.1.1 Overview of Machine Learning for Resource Management

ML allows systems to respond better to changing needs by recognizing patterns and predicting workloads. The system helps decide wisely without step-by-step directions. It handles needs when wants change in unexpected ways. Methods that learn are very useful for times with shifting, unclear duties.

#### 10.6.1.2 Predictive Resource

Predicting how much is needed helps machines use things wisely. By looking at what was used before and what work needs to get done now, algorithms can guess what will be needed later. This smart way of preparing cuts down on wasting time waiting and uses what's available in the best way. It lets systems get or return what might be needed before needs change.

#### 10.6.1.3 Fault Prediction and Anomaly Detection

Early detection of anomalies allows systems to run smoothly by stopping actions like moving resources or self-fixing. Distributing resources flexibly is possible using algorithms that learn and adjust assignments as patterns change. To best share resources, these algorithms consider how well applications work, how people use them, and outside conditions. This versatility makes sure resources meet each service's and app's unique needs. Reinforcement learning creates smart rules. These algorithms build the top resource strategies through tests and mistakes. The rules can change to new situations, permitting the framework to improve over time as trends and issues emerge or change.

## 10.6.2 Methods of Optimisation

### 10.6.2.1 Introducing Resource Orchestration's Optimisation Techniques

The aim of optimization methods is to find the right way to share and use things among connected systems. To get the wanted results, these calculating and math processes make the best of some things like cost, energy use, and how fast work gets done. Problems of giving out what's needed are written as formulas and solved using both straight and bendy math. These models think about goals and limits, letting what's needed be given out as good as can be while meeting what was set first. Bendy math deals with harder ties, while straight math works for things tied straight.

Biological evolution inspires evolutionary computing and genetic algorithms to find quality solutions. A group of workable solutions is made, their effectiveness is judged, and the group gradually improves to discover the top solution. Genetic algorithms can examine different resource distributions in resource orchestration to pinpoint the most beneficial arrangement. By decreasing wait times and boosting how many are helped, queuing theory aids optimize resource sharing by acting out how activities or processes flow within a system. Orchestration systems can assist the use of computing power better by distributing resources based on anticipated workloads and prioritizing important processes by studying the ebbs and flows of lines.

Game theory can help with managing shared things when different people or groups are involved. To be most helpful, each person thinks carefully about what others might do before deciding their own plan. Using game theory, especially for computers working together makes sure everything balances between what each person wants and what's best for everyone sharing the resources.

## 10.6.3 Hybrid Models

### 10.6.3.1 Optimisation Through Machine Learning Hybrids

ML predictions and optimization methods together form hybrid models. This helps decide how to use things in the best way. For example, optimization can find the best way to share work based on what ML expects might happen later. Working together makes everything flexible and smart.

### 10.6.3.2 Combining Rule-Based and Learning-Based Methods: Advancing Hybrid Approaches

Some models blend ML with usual rule-based methods. Fixed policies follow guidelines, but changing situations are discovered and adjusted by learning algorithms. This mix lets people build a flexible coordination system that obeys set standards yet modifies when novel information patterns appear.

#### 10.6.3.3 Continual Enhancement Through Responsive Feedback Mechanisms

Feedback loops let hybrid models continuously upgrade tactics. For example, real-time info helps AI programs learn. Optimization plans adjust how things are shared depending on shifting targets. Decision usefulness is checked through feedback systems too. The way everything works together is sure to evolve and alter gradually thanks to this step-by-step way.

#### 10.6.3.4 Harnessing the Power of Adaptive Model Switching

At times the system chooses one method over another. For example, when things are unpredictable, it relies more on what it has learned from data. When conditions are steady, it uses formulas to make choices. By switching between the two approaches, it ensures great results in any circumstances.

## 10.7 Case Studies

Flexible planning helps computing work best, as shown by many real programs. They use resources in smart ways, do faster, and change as needs change. This part looks at real programs that did well and the useful things we learned from them.

### 10.7.1 Practical Applications

#### 10.7.1.1 AWS

- **Implementation:** Auto scaling analyzes previous patterns to predict future needs for resources. It uses past data to forecast demand, allowing

#### 10.7.1.2 Autoscaling of Kubernetes Horizontal Pods

- Essentially, HPA ensures the application environment is exactly set by adapting to changing workloads by automatically altering the number of pods to use resources to their fullest.
- **Impact:** Kubernetes HPA users benefit from heightened resource reactiveness and productivity, which guarantees that apps may automatically change in response to changing demand.

## 10.7.2 Achievements and Insights Acquired

### 10.7.2.1 Netflix: Using Machine Learning to Deliver Content

- Netflix, one of the top streaming services, uses smart programs to make streaming run smoothly. They watch how people use the site, what shows are popular, and how busy the internet is. Then the programs adjust the delivery of shows and movies. This helps Netflix deliver content without interruptions. By paying attention to viewers and the network, their system makes streaming work well.
- **Lessons Learned:** Netflix has shown that using numbers and facts to guide choices can help a business do well. The service can shift what it offers quickly as people's interests change or the internet acts up. It does this with programs that learn on their own.

### 10.7.2.2 Google's Expansion of Kubernetes: Enhancing Scalability

- Google, the creators of Kubernetes, have shown how useful the tool can be. They apply it throughout their huge networks to coordinate programs split into portable parts. This ensures computer power and memory are shared smoothly and flexibly between tasks. No matter the workload, Kubernetes keeps resources assigned properly so their systems keep performing.
- Google's time utilizing Kubernetes demonstrated how container direction can expertly deal with perplexing and expansive registering frameworks. A portion of the significant lessons are how standardized administration structures are important, and how element asset allotment benefits a condition intended for the cloud.

### 10.7.2.3 Achieving Dynamic Scalability with AWS Auto Scaling: An Airbnb Success Story

- Airbnb has become very popular around the world. People can find places to stay when they travel using the Airbnb website and app. Airbnb uses Amazon Web Services (AWS) to help their computers work better. AWS helps Airbnb's computers change how many are working based on how many people are using the site. This makes sure enough computers are working when lots of people are booking rooms. Airbnb set up automatic rules to make the computers change fast.

Our learning: The importance of cloud-based automatic resizing was shown by how Airbnb used AWS auto scaling. We have discussed various lessons such as handling big website traffic efficiently, saving money by resizing alone, and having the freedom to share resources in different ways.

## 10.8 Conclusion

Orchestrating computing, data, and internet-connected devices (such as IoT) is important for businesses using networks. Looking closely at how resources are used, a new way is emerging. Instead of keeping resources the same all the time, adapting them based on current needs works better. This DRO is changing how IT systems are designed. The old way of setting resources and leaving them fixed is being replaced with one that adjusts computing power, data storage, and IoT devices flexibly. Dynamic orchestration uses ML, optimization methods, and mixed approaches. Figuring out complex network demands and optimizing allocations while adapting to changes requires sophistication. Companies can benefit too. From these systems, important lessons appear. Data-driven decisions, common orchestration frameworks, and adjusting resources dynamically can improve efficiency. As networks add edge computing, more connected devices, and advanced analysis, DRO plays an even greater role. It keeps IT environments stable, safe, and adaptable to changing technology needs and resource usage.

## References


1 Guim, F., Metsch, T., Moustafa, H. et al. (2022). Autonomous lifecycle management for resource-efficient workload orchestration for green edge computing. *IEEE Transactions on Green Communications and Networking* 6 (1): 571–582. https://doi.org/10.1109/TGCN.2021.3127531.

2 Dona, M.A.M., Berger, C., and Yu, Y. (2023). AirDnD - asynchronous in-range dynamic and distributed network orchestration framework. *2023 IEEE 43rd International Conference on Distributed Computing Systems (ICDCS)*, Hong Kong, Hong Kong (18-21 July 2023). IEEE, pp. 953–954, https://doi.org/10.1109/ICDCS57875.2023.00108.

3 Sattari, A., Ehsani, R., Leppänen, T. et al. (2020). Edge-supported microservice-based resource discovery for mist computing. *2020 IEEE Intl Conf on Dependable, Autonomic and Secure Computing, Intl Conf on Pervasive Intelligence and Computing, Intl Conf on Cloud and Big Data Computing, Intl Conf on Cyber Science and Technology Congress (DASC/PiCom/CBDCom/CyberSciTech)*, Calgary, AB, Canada (17-22 August 2020). IEEE, pp. 462–468, https://doi.org/10.1109/DASC-PICom-CBDCom-CyberSciTech49142.2020.00087.

4 Katenbrink, F., Seitz, A., Mittermeier, L. et al. (2018). Dynamic scheduling for seamless computing. *2018 IEEE 8th International Symposium on Cloud and Service Computing (SC2), Paris, France* (18-21 November 2018). IEEE, pp. 41–48, https://doi.org/10.1109/SC2.2018.00013.

**5** Lam, A.N., Haugen, O., and Delsing, J. (2022). Dynamical orchestration and configuration Services in Industrial IoT systems: an autonomic approach. *IEEE Open Journal of the Industrial Electronics Society* 3: 128–145.

**6** Imrith, V.N., Ranaweera, P., Jugurnauth, R.A., and Liyanage, M. (2020). Dynamic orchestration of security services at fog nodes for 5G IoT. *ICC 2020–2020 IEEE International Conference on Communications (ICC)*, Dublin, Ireland (07-11 June 2020). IEEE, pp. 1–6.

**7** Ji, L., He, S., Wu, W. et al. (2022). Dynamic network slicing orchestration for remote adaptation and configuration in industrial IoT. *IEEE Transactions on Industrial Informatics* 18 (6): 4297–4307.

**8** Imrith, V.N., Ranaweera, P., Damree, S., and Liyanage, M. (2021). Enabling fog computing based dynamic security service function chaining for 5G IoT. *2021 IEEE International Conference on Advanced Networks and Telecommunications Systems (ANTS)*, Hyderabad, India (13-16 December 2021). IEEE, pp. 149–154.

**9** Hakiri, A., Sellami, B., Ben Yahia, S., and Berthou, P. (2020). A SDN-based IoT architecture framework for efficient energy management in smart buildings. *2020 Global Information Infrast and Networking SymposGIIS)*, Tunis, Tunisia (28-30 October 2020). IEEE, pp. 1–6.

**10** Alhameed, M. and Hossain, M.A. (2023). Rapid detection of pilgrims whereabouts during hajj and umrah by wireless communication framework: an application AI and deep learning. *2023 International Conference on Emerging Smart Computing and Informatics (ESCI)*, Pune, India (01-03 March 2023). IEEE, pp. 1–6, https://doi.org/10.1109/ESCI56872.2023.10099969.

**11** Refaee, E.A., Hossain, M.A., Soundrapandiyan, R. et al. (2022). Biomedical image retrieval using adaptive neuro-fuzzy optimized classifier system. *Mathematical Biosciences and Engineering: MBE* 19 (8): 8132–8151. https://doi.org/10.3934/mbe.2022380. PMID: 35801460.

**12** Alhameed, M., Jeribi, F., Elnaim, B.M.E. et al. (2023). Pandemic disease detection through wireless communication using infrared image based on deep learning [J]. *Mathematical Biosciences and Engineering* 20 (1): 1083–1105. https://doi.org/10.3934/mbe.2023050.

**13** Buzachis, A. and Villari, M. (2018). Basic principles of osmotic computing: secure and dependable microElements (MELs) orchestration leveraging blockchain facilities. *2018 IEEE/ACM International Conference on Utility and Cloud Computing Companion (UCC Companion)*, Zurich, Switzerland (17-20 December 2018). IEEE, pp. 47–52, https://doi.org/10.1109/UCC-Companion.2018.00033.

**14** Ruggeri, G., Amadeo, M., Campolo, C. et al. (2021). Caching popular transient IoT contents in an SDN-based edge infrastructure. *IEEE Transactions on Network and Service Management* 18 (3): 3432–3447. https://doi.org/10.1109/TNSM.2021.3056891.

**15** Tariq, M.A., Farooq, M.U., Zeeshan, M. et al. (2022). FIPAM: fuzzy inference based placement and migration approach for NFV-based IoTs. *IEEE Transactions on Network and Service Management* 19 (4): 4298–4309. https://doi.org/10.1109/TNSM.2022.3145658.

**16** Amjad, S., Akhtar, A., Ali, M. et al. Orchestration and management of Adaptive IoT-centric distributed applications. *IEEE Internet of Things Journal* https://doi.org/10.1109/JIOT.2023.3306238.

**17** Porambage, P., Okwuibe, J., Liyanage, M. et al. (2018). Survey on multi-access edge computing for internet of things realization. *IEEE Communications Surveys & Tutorials* 20 (4): 2961–2991.

**18** Verma, S., Kawamoto, Y., Fadlullah, Z.M. et al. (2017). A survey on network methodologies for real-time analytics of massive IoT data and open research issues. *IEEE Communication Surveys and Tutorials* 19 (3): 1457–1477.

**19** Liyanage, M., Ahmad, I., Abro, A.B. et al. (2018). *A Comprehensive Guide to 5G Security*. John Wiley & Sons.

**20** Quick, D. and Choo, K.-K.R. (2018). IoT device forensics and data reduction. *IEEE Access* 6: 47566–47574.

**21** Liyanage, M., Braeken, A., Kumar, P., and Ylianttila, M. (2020). *IoT Security: Advances in Authentication*. John Wiley & Sons.

**22** Kolias, C., Kambourakis, G., Stavrou, A., and Voas, J. (2017). DDoS in the IoT: Mirai and other botnets. *Computer* 50 (7): 80–84.

**23** Ranaweera, P., Jurcut, A.D., and Liyanage, M. (2019). Realizing multi-access edge computing feasibility: security perspective. *2019 IEEE Conference on Standards for Communications and Networking (CSCN)*, Granada, Spain (28-30 October 2019). IEEE, pp. 1–7.

**24** Zhang, P., Liu, J.K., Yu, F.R. et al. (2018). A survey on access control in fog computing. *IEEE Communications Magazine* 56 (2): 144–149.

**25** Aazam, M., Zeadally, S., and Harras, K.A. (2018). Fog computing architecture evaluation and future research directions. *IEEE Communications Magazine* 56 (5): 46–52.

**26** Ismail, B.I., Goortani, E.M., Ab Karim, M.B. et al. (2015). Evaluation of docker as edge computing platform. *2015 IEEE Conference on Open Systems (ICOS)*, Melaka, Malaysia (24-26 August 2015). IEEE, pp. 130–135.

**27** Boudi, A., Farris, I., Bagaa, M., and Taleb, T. (2019). Assessing lightweight virtualization for security-as-a-service at the network edge. *IEICE Transactions on Communications* 102 (5): 970–977.

**28** Wong, K., Dillabaugh, C., Seddigh, N., and Nandy, B. (2017) Enhancing Suricata intrusion detection system for cyber security in SCADA networks. *2017 IEEE 30th Canadian Conference on Electrical and Computer Engineering (CCECE)*, Windsor, ON, Canada (30 April 2017 - 03 May 2017). IEEE, pp. 1–5.

**29** Park, W. and Ahn, S. Performance comparison and detection analysis in snort and suricata environment. *Wireless Personal Communications* 94 (2): 241–252.

# 11

# Cognitive Cellular Networks: Empowering Future Connectivity Through Artificial Intelligence

*Mohammad Alamgir Hossain[1], Suresh Limkar[2], Sherif Tawfik Amin[1], and Yasir Ahmad[3]*

[1]*Department of Computer Science, College of Computer Science & information Technology, Jazan University, Jazan, Kingdom of Saudi Arabia*
[2]*Department of Artificial Intelligence & Data Science, AISSMS Institute of Information Technology, Pune, Maharashtra, India*
[3]*Department of Information Technology & Security, College of Computer Science & Information Technology, Jazan University, Jazan, Kingdom of Saudi Arabia*

## 11.1 Introduction

### 11.1.1 Background

The connection between mobile networks and machine learning marks an important change in how communication technology may grow, with big effects on how we find and share information. This ground-breaking team-up is shown by the rise of smart cell networks, a big change in how communication systems work. Merging AI codes with regular cell networks makes them stronger, use what they have better, and adjust fast to [1] what people will need.

In the past, cell phone networks have been the main way people connect with each other all over the world. However, now AI is transforming [2, 3] cellular networks. Machine learning uses past information to predict what people will do. Networks can plan ahead using these predictions. They can get ready for spikes in usage and make sure everyone has enough bandwidth. Problems get fixed before they start. AI helps networks think ahead and meet our needs better than before [4].

Machine learning helps computers study what happened before to guess what may happen next. By looking at old information carefully, AI programs can help networks make choices before problems happen. This lets networks use things like bandwidth and servers better. It also lets networks change quickly as more apps

*Current and Future Cellular Systems: Technologies, Applications, and Challenges*, First Edition.
Edited by Garima Chopra, Suhaib Ahmed, and Shalli Rani.

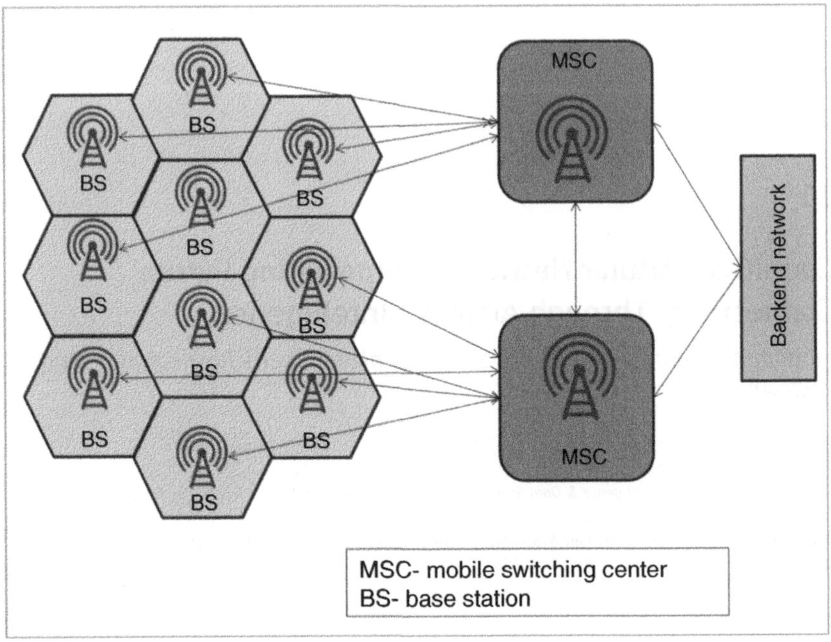


**Figure 11.1** Overview of cellular network architecture.

and devices join the Internet. Acting before issues arise boosts how well networks serve people and keep up as technology advances. Furthermore, AI's ability to analyze goes beyond normal limits. It can spot hidden rules and odd things within network actions, making security stronger and lowering the chances of electronic crimes working [5, 6]. This results in a communication setup that is protected against possible dangers, so it guarantees sharing and getting data is safer. So, using AI does not just improve how communication systems function but also builds a sturdy barrier against the ever-changing troubles caused by the digital world (Figure 11.1).

### 11.1.2 Key Objectives of the Chapter

- Examine the ways in which machine learning algorithms might improve prediction abilities, adjust to changing communication settings, and optimize network characteristics.
- **Application Insights:** Examine the wide range of applications enabled by CCNs in a number of industries, such as manufacturing, entertainment, transportation, and healthcare.
- **User Experience Enhancement:** Research how new types of networks influence how people use technology, focusing on reduced delays, reliable connections, and working with developing tech like augmented reality. Study how

artificial intelligence (AI) and wireless carriers impact how individuals connect with digital tools.

- Think about how adding AI to mobile networks could change security. Some long sentences: Machine learning has the potential to boost [7] protection by closely watching for vulnerabilities, catching attempts to break in, and getting smarter over time to better safeguard connectivity.

This study aims to fully explain cognitive cellular networks (CCNs). It will look at the hard challenges they present technically. It will also look at the different ways they can be used. The study wants to make them better for people to use. It will also deal with the important issue of keeping them safe online. The goal is to add to what is being talked about now. That is about how AI could change phone networks and how we connect in big ways.

## 11.2 Foundations of Cognitive Cellular Networks

### 11.2.1 Architecture of Cellular Networks

Cellular networks have a complex design that has changed significantly from first-generation to fifth-generation networks and beyond. This complex design is the basis for CCNs [8]. Traditional cellular networks consist of devices and towers in a hierarchy. These conventional setups have struggled to keep up with growing data traffic and different app needs. Fifth-generation networks will allow very fast speeds and low delays thanks to a more spread out and decentralized setup using massive multiple input multiple output (MIMO) systems and small cells. Cognitive functions are added to cellular network designs in CCNs. This enhances networks even more. An adaptable system using AI optimizes several aspects of performance instead of traditional fixed configurations. AI algorithms are responsible for smartly placing and managing small cells. They help identify the network and boost its capacity. CCNs can instantly react to changing user demands because of their flexible structure. It guarantees effective resource use and better user experiences.

### 11.2.2 Radio Technologies Induced by Cognition

A crucial part of putting CCNs into practice involves cognitive radio technologies (CRTs), which allow flexible use of radio waves, as shown in the diagram [9]. Historically, specific radio frequencies were set aside permanently for each communication channel. However, this led to wasted airwaves and inefficient use of the spectrum. CRT introduces the idea of cognitive radios. These radios can sense shifts in how the spectrum is being used, alter their settings accordingly, and make good use of empty frequencies.

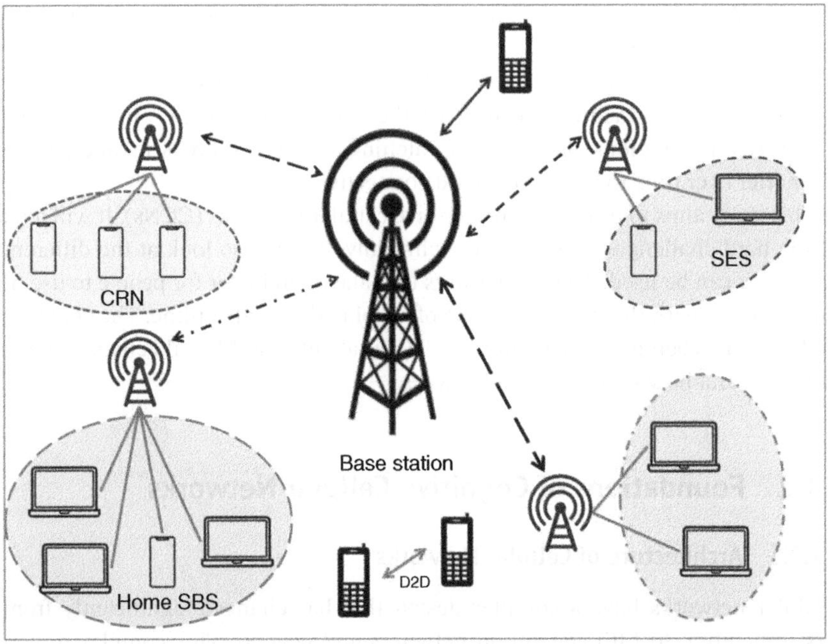


**Figure 11.2** Representation cognitive radio technology in communication.

When discussing cellular [10] cognitive networks, cognitive radio technology is extremely important for making the best use of available radio waves. Smart radios can flexibly pick the perfect channels based on current surroundings and cleverly change frequencies across open bands with the help of AI programs. This adjustment of spectrum usage decreases issues with signal disruption and network overload while upgrading the mobile network's all-around performance [11]. What's more, CRT in CCNs encourages better dividing up of spectrum and a more helpful and adaptable means of managing available frequencies (Figure 11.2).

### 11.2.3 Artificial Intelligence Integration

AI [6] is smoothly merged with CCNs, providing the network structure with intellectual abilities. AI functions as CCNs' minds, allowing them to study, change, and make sensible conclusions independently. A part of AI known as machine learning formulas is crucial for changing cellular networks into conscious structures that can optimize functionality in several circumstances [12].

Predictive analytics uses AI to foresee how folks will act with applications, what they may need, and expected network conditions. Machine learning models study past network [13] records to make this prediction. Moreover [14], AI-powered analysis helps improve security. By studying traffic patterns to spot abnormalities, machine learning identifies threats. Protecting confidentiality and integrity within the cellular network requires identifying risks ahead of time.

The mix of advanced cell [15] tower design, radio technologies that can think, and the smooth addition of AI is the foundation for CCNs. This joining opens doors for clever, adjustable, and good communication systems besides solving the hard things caused by cellular networks needing to do more and more. As we discover more about how complicated CCNs are, fitting in AI more and more is obvious as a revolutionary power that is taking cell networks into a new time of unbeaten smartness and reactions.

## 11.3 AI Algorithms for Network Optimization

### 11.3.1 Machine Learning Models for Predictive Analysis

#### 11.3.1.1 Machine Learning in Resource Allocation

How computers share things is important. Two kinds of machine learning are often used. One is called reward learning. The other is called deep learning. They help computers share things like bandwidth in fair and changing ways. These models look at how networks and people act. They also look at what happened before. This helps predict what will happen next and share things where needed. When computers share things this way, it saves bandwidth and makes things faster. Users also get better service overall.

#### 11.3.1.2 Predictive Analytics for Traffic Management

Predicting network demands and distributing traffic is important for maintaining a good CCN. Machine learning is used to analyze past network use and estimate future needs. The algorithms help anticipate when more people will be online, what apps they will use, and how users will act. This [16] allows the network to prepare ahead of time. It can adjust before traffic gets too crowded and users have issues. Because the AI spreads [17] usage evenly over different parts of the network, predicting what will happen helps balance the loads. Keeping traffic steady in all areas prevents any one spot from slowing down.

#### 11.3.1.3 Reinforcement Learning for Self-Optimizing Networks

The area of reinforcement learning (RL) has fantastic potential to enable self-optimizing systems in communication networks. Cell networks can learn from their experiences with conditions around them and make choices that lead to the best outcomes due to RL algorithms. RL [18] is used in communication networks for jobs like adjusting settings without guidance, which alters network qualities in real time based on how well things are working feedback, like how strongly signals are sent, when to connect to a new tower, and sharing resources. The network works better and reacts faster due to this automatic adjustment, particularly in changing and unpredicted situations involving sending messages.

#### 11.3.1.4 Anomaly Detection to Strengthen Security

AI methods using irregularity identification strategies vastly improve CCN protection. To notice variances from what's standard, machine learning designs containing self-supervised studying formulas examine system traffic designs. Oddities could signify unauthorized access attempts or denial-of-service assaults, two possible protection dangers. CCNs with irregularity identification formulas can rapidly pinpoint and handle protection risks by constantly tracking and assessing system knowledge [19]. This enhances the interaction framework's total cybersecurity.

#### 11.3.1.5 Artificial Neural Networks for Dynamic Optimization

Swarm intelligence algorithms are used in dynamic optimization [20] within CCNs and are inspired by collective behavior seen in nature. To optimize network characteristics, algorithms like ant colony optimization and particle swarm optimization mimic the cooperative decision-making processes of swarms. These algorithms work especially well in situations where it is necessary to adjust several parameters at once. Swarm intelligence has a role in energy-efficient routing, dynamic frequency allocation, and adaptive antenna layouts in CCNs.

#### 11.3.1.6 Combining Genetic Algorithms with Cross-Layer Optimization

Natural selection serves as the inspiration for genetic algorithms, which are used in CCNs for cross-layer optimization. By [21] taking into account interactions between various network stack layers, these algorithms optimize parameters in a coordinated way.

### 11.3.2 Spectrum Utilization and Management

CCNs use radio waves in different ways than normal cell networks. They are made to meet all the needs of different apps now and in the future. Old cell networks don't use radio waves well and don't have enough frequencies. CCNs change how frequencies are shared using smart tech and new ideas. They make sure all the available radio waves are used in the best way. They can also adjust how things work based on what is happening around them.

#### 11.3.2.1 Dynamic Spectrum Access

DSA [22] allows communication networks to intelligently use available radio frequencies depending on current needs and environmental conditions, unlike set frequency assignments in traditional networks. This is particularly important considering the limited and valuable radio waves.

#### 11.3.2.2 Brain CRT

These cognitive radios contain AI programs allowing them to optimize how they act over time and decide alone about the available spectrum. Because CRT makes flexible spectrum access easier, CCNs can move around the spectrum dynamically and take advantage of unused frequency areas. As a result, the spectrum is used more effectively with less interference, and other wireless systems get along better. Cognitive radios improve the general adaptability and resilience of CCNs in changing and busy environments by continuously learning from their surroundings [23, 24].

#### 11.3.2.3 Enhancing Spectrum Management with AI-Powered Solutions to Combat Interference

Cellular networks regularly faced disruptions that harmed how people used the service and how well it worked. Networks now use smart technology powered by AI to better control interference. Machine learning [25–29] programs can adjust how the signals are sent, find what is causing problems, and study how the interference happens. Networks also share local spectrum information between devices using collaborative sensing helped by AI. Working together allows more accurate decisions about airwaves, lowering the chance of interference. The result is that the network continuously aims to reduce disturbances from interference while also making the best use of what spectrum is available (Figure 11.3).

#### 11.3.2.4 Achieving Regulatory Compliance in Spectrum Sharing

Spectrum sharing and regulatory compliance are supported through smart algorithms in CCNs. The networks can optimize how they use radio waves while following the law by constantly checking use and changing things to obey rules. AI also helps different groups share airspace better and more smoothly, promoting teamwork and flexibility when managing frequencies [30].

## 11.4 Reinforcement Learning in Autonomous Network Management

CCNs are changing how networks work. They use a type of machine learning called RL. This helps networks manage themselves without people. RL teaches networks to make choices on their own based on what worked before. It helps networks adjust as things change. Some networks now use RL algorithms. Studies show these algorithms can manage networks well without supervision. They help networks operate independently in changing situations. This section

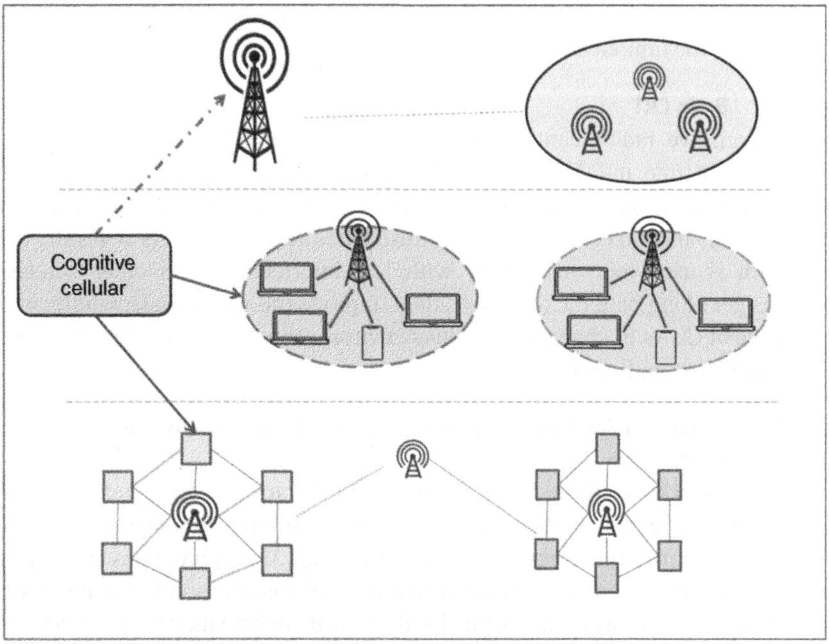


**Figure 11.3** Representation of cognitive spectrum sharing.

will look at RL basics. It will also show how networks can adapt automatically over time. Examples will demonstrate how RL algorithms help networks run themselves.

### 11.4.1 Essential Guidelines for Mastering Reinforcement Learning

A branch of machine learning called RL focuses on the idea of something interacting with its surroundings. When managing networks automatically in CCNs, the surroundings refer to the complex and constantly changing conditions in the network. The thing interacting could be a network node or the controller. In these surroundings, the thing acts and gets feedback in the form of rewards or punishments based on how its actions turn out. The possible situations, options available, rules followed, how actions are measured now, and how valuable options are later are key parts of RL. The options available consist of the choices open to the thing acting. The possible situations show different setups or scenarios that could happen in the surroundings. The rules that guide the thing's decision-making are called the policy. And the reward function measures how good an action is right now (Figure 11.4).

Some situations are better than others over the long run. RL lets things teach themselves in networks by trying different choices and seeing what helps the most.

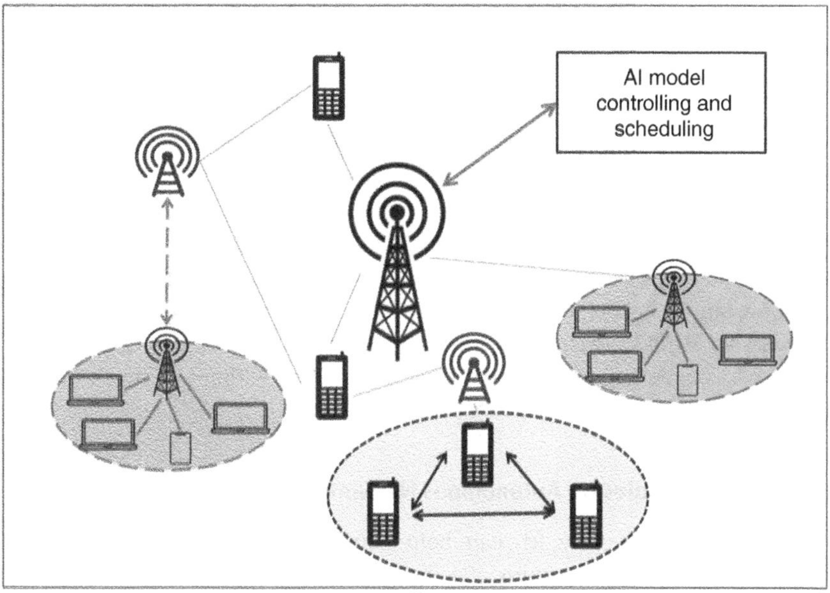


**Figure 11.4** Dynamic AI modeling for cellular network.

They explore options and learn step-by-step which choices work best. The things adjust what they do to get the most help, adding up through this learning as they go. This makes the network run better and uses resources wisely.

## 11.4.2 Adaptive Decision-Making in Dynamic Environments

RL's [31] ability to make choices that adapt to shifting scenarios defines it. The adaptability of self-guided choices proves important for computer networks connecting people since how people act, the network works, and what software needs to alter constantly.

### 11.4.2.1 Time-Based Learning and the Trade-Off Between Exploration and Exploitation

A key way for computer programs to learn over time is called temporal difference learning. It allows programs to change what they know based on the difference between what reward they expect and what reward they get. This is very helpful for programs working in networks that often change. It lets the programs in networks adjust their plans in response to changing conditions in the network. Also, how RL works ensures the programs balance trying new things to maybe find better strategies by using methods they already know well to get quick results.

#### 11.4.2.2 Dynamic Approaches to State Representation and Policy Adaptation

Creating policies that respond to the network's current state is key for RL in communication networks. A crucial part is how the network status is represented. Over time, RL agents refine their policies to make informed choices by considering traffic, latency, and user needs. In these networks, RL is commonly used for adjusting resources in real time. Stations and controllers act as RL agents trained to allocate electricity, bandwidth, and computing power based on changing traffic. This ensures the network satisfied different services and users when demand fluctuates. Networks can optimize parameters on their own thanks to RL. For example, agents in a 5G network might adjust beam-forming parameters in a massive MIMO system to improve signals and coverage. Self-organization is supported as agents autonomously refine parameters without as much manual work, boosting the whole network's performance [32].

### 11.4.3 Case Studies on Autonomous Network Management

This paper looks at how RL can help networks run themselves. It shows examples of how RL algorithms effectively fixed real problems in content networks. The algorithms helped the networks learn from what worked and what did not.

**Case Study 11.1**

**5G Networks Using Dynamic Spectrum Access**

Using airwaves differently helps networks run well in 5G and beyond. Base stations and smart radios may find the best ways to share airwaves based on where they are. They learn on their own thanks to programs that reward good choices. The programs help agents change how they access airwaves based on lots of things like interference, how many people use the network, and rules. This leads to using airwaves better and fewer things blocking signals.

**Case Study 11.2**

**Self-Configuring Small Cell Placements**

Making cellular networks denser to boost how much they can handle and reach is an important job for small cells. With assistance from algorithms that learn, tiny cells may set up their own placements, accounting for how the network is at that time. To keep connections smooth and share the workload evenly when things change, devices are trained to adjust how strongly they broadcast, the angle of antennas, and when to hand over users. This ability to configure itself adds to cellular core networks' strength overall in times when how many users there are goes up and down.

**Case Study 11.3**

**Pre-emptive Identification and Reduction of Faults**

Fixing problems before they happen is important for networks. Agents learn to spot signs that could cause issues by always looking at how well the network is doing. They find mistakes and fix them themselves by sending traffic other ways or changing settings to help. Finding and fixing faults early makes networks stronger and better able to deal with challenges. This helps networks work well for people.

**Case Study 11.4**

**Mobile Edge Computing (MEC) Quality of Service (QoS) Optimization**

RL algorithms help make MEC better. They can make services work better for people using phones, tablets, and other devices. Agents learn how to choose the best ways to share tasks between devices and clouds. They keep work balanced and resources available where needed. People accessing edge services get fast and smooth experiences. This happens because the agents keep making choices based on what is needed at that moment.

**Case Study 11.5**

**mMIMO Systems with Adaptive Beamforming**

RL helps huge systems with many antennas at both ends figure out the best ways to aim their signals. RL helpers learn the best techniques for sending beams of radio waves to different devices at once. This boosts how much data the system can handle at a time. It also reduces messy signals getting in each other's way. The beams

## 11.5 Applications of Cognitive Cellular Networks

CCNs merge artificial thinking with mobile networks in new ways. CCNs can reshape how people see and use being connected. With focuses on faster mobile

Internet, many devices talking, and ultra-reliable instant connections, this part looks at CCN applications. It also explores specific uses and real examples showing how CCNs can revolutionize many fields.

### 11.5.1 Upgraded Mobile Broadband

CCNs help improve mobile Internet access, especially with 5G and future networks. The network uses smarts to best manage its resources, change how much bandwidth devices get, and make mobile Internet better. This makes data transfer much quicker, network capacity much greater, and spectral efficiency much higher.

- **Maximizing Spectrum Efficiency Through Dynamic Utilization**
  Enhanced mobile broadband (eMBB) can use different ways to use radio waves thanks to CCNs being better at managing how the radio waves are used. Machine learning techniques make bandwidth sharing better and automatically change what frequencies are used by looking at what is happening with the radio waves now, what people want, and how busy the network is. By making sure eMBB services have what they need, this changing plan leads to faster data speeds and a mobile broadband service that reacts better.
- **Enhancing Data Speeds with Adaptive Beamforming**
  Beamforming technology powered by CCNs is crucial for boosting signal quality and throughput in eMBB applications. With RL algorithms providing direction, large antenna systems continually adjust their configurations to focus signals precisely where users need them most. This adaptability elevates the general standard of mobile Internet by permitting quicker downloads, reducing disruptions, and extending reach.

### 11.5.2 Massive Machine-Type Communication

Different devices need to connect in large numbers. Making lots of things talk to each other is the main goal of the Internet of Things (IoT). How networks handle different amounts of devices is important. Networks must meet the needs of all kinds of devices and how much they communicate. CCNs can change to fit different device densities and conversation requirements. This flexibility is key for applications where many machines must talk.

- **Enhancing Connections Through Dynamic Management**
  At times, many tools talking makes the network busy. Networks use computer learning to change how they connect tools. The network can change to deal with tools sending bits at odd times or a lot quickly. Because it can bend, the network shares its parts well. This slows how much data moves and helps tools talk better even if they use the network in different ways.

- **Energy-Efficient Communication**
  Connected devices have limited power, so using energy wisely is important. AI helps communication networks save device battery power. By learning how devices act, the network can change when and how they talk to each other. These small changes help devices last longer without making communication unreliable.

### 11.5.3 Ultra-reliable Low-Latency Communication

Some programs need fast responses and steady work, like virtual reality glasses, factory machines, and driverless cars. They count on quick data sharing. CCNs fill the tight needs of ultra-reliable low-latency communications (URLLC) by using smarts for new adjustments and choices.

- **Achieving Real-Time Autonomy: Empowering Swift Decision-Making**
  In time-crucial network settings, smart connection networks use the extra machine-thinking strategies like RL to make independent conclusions. In industrial machine work, as an example, the system can self-change association settings to ensure ultra-low time delay, permitting fundamental tasks to be overseen in genuine time. For uses where even milliseconds of time delay may have enormous outcomes, this flexibility is fundamental.
  Predictive analytics for URLLC applications is aided by machine learning techniques found in CCNs. This leads to predictive analytics for reliability. The network can proactively address dependability concerns by examining historical data and foreseeing possible problems. By taking a proactive stance, the likelihood of communication breakdowns is reduced, providing mission-critical applications with a highly dependable and low-latency communication environment.

### 11.5.4 Use Cases and Practical Implementations

The revolutionary power of CCNs is demonstrated through concrete use cases and real-world applications in a variety of industries, going beyond theoretical applications.

a) Healthcare
   By enabling telemedicine applications and remote patient monitoring, CCNs transform healthcare. Dynamic spectrum utilization and adaptive beamforming are beneficial for eMBB real-time data transfer from wearable devices and high-quality video consultations. Reliable communication is ensured in massive machine-type communication (mMTC) by the network's efficient handling of the huge influx of data from healthcare IoT devices. Low-latency

communication is ensured by URLLC for vital health monitoring applications, improving patient care.

b) Smart Cities
CCNs support effective public safety, infrastructure monitoring, and traffic control in the context of smart cities. High-speed connectivity for homes, businesses, and IoT devices is guaranteed by eMBB.

## 11.6 Challenges and Future Directions

### 11.6.1 Scalability and Standardization

How well networks can scale and the consistency of standards are closely tied. Lacking uniform standards may cause issues with things working together and make integrating the networks into existing communication setups more difficult. To build a structure supporting things working together and scalability, those in industries, groups setting standards, and oversight groups must collaborate. By solving these issues, networks connecting computers, people, and machines will be widely used and a globally connected system encouraged.

### 11.6.2 Future Trends in Cognitive Cellular Networks

Future CCNs are expected to be shaped by a number of trends that will likely push them in the direction of increased intelligence, flexibility, and efficiency.

- **AI-Driven Edge Computing:** This new idea for computer networks combines AI with edge computing. Networks can give users better time and faster answers by making computers smarter near people and their devices at the edge of the network. By helping apps like augmented reality, Internet-connected machines, and self-driving cars make choices right away, edge AI helps build a communication setup that is quicker and more reactive.
- **5G and Beyond Integration:** As the abilities of both technologies grow, the connection between CCNs and 5G systems will become even more powerful. CCN combining with 5G and advancing beyond 5G (B5G) innovations will unlock brand new chances for rapid interaction, massive device assistance, and enhanced linking. To completely make use of smart and high-performing communication networks, this blending is critical.
- **Explainable AI for Network Management:** Providing clarity on AI decisions in networks is growing more necessary as AI systems for communications become more intricate. Crafting AI that gives an obvious understanding of its resolution-making is a developing pattern. In networked communication, AI with transparent rationales will not just boost user certainty in self-managing

frameworks but also enhance cooperation between AI and human personnel, particularly when crucial choices must be made.

- **AI for Predictive Network Maintenance:** Networks will need to prepare for the future. Looking at past data can help find issues before they cause problems. Computers will get better at noticing small changes and patterns that suggest troubles may happen soon. These smart systems will watch how the network works over time. They can learn to see when something might stop working in the near future. This predictive technology will make networks more steady and reliable. It can cut down on how often the network goes down. It will also lessen the burden on the people who maintain the network's operations.
- **Self-Learning Networks:** The concept of networks that learn by themselves, where CCNs consistently change and enhance how they work in response to suggestions and what they learn directly, is anticipated to become more common. By using adjustable AI strategies like RL, networks will have the ability to learn from changing conditions, how people act, and fresh applications. Possessing this ability to learn by themselves will help develop networks that are extremely aware of changing communication contexts and smart too.

The future holds exciting possibilities along with issues that must be addressed for CCNs to connect seamlessly. Security, size, and shared standards need solutions. The merging of smart tech and mobile networks will transform how we link, talk, and use technology. Researchers, companies, and regulators must work together to handle the complex details of evolving CCNs and access their full power to make networks intelligent, adjustable, lasting, and safe. CCNs' path will surely shape the digital world, encouraging a linked future fitted to the different needs of the digital era.

## 11.7 Conclusion

Cognitive cellular networks, or CCNs, are transforming communication technology by integrating AI into standard cellular systems. Advanced AI methods such as machine learning and RL are applied within CCNs to provide smart services that adapt to changing conditions and optimize the use of resources. CCNs can address a variety of needs effectively, as seen in eMBB, massive machine-type communication, and ultra-reliable low-latency communication. CCNs may revolutionize many industries and improve lives in different ways. For example, they could enhance city infrastructure and enable real-time connections for driverless cars. CCNs could also transform healthcare through remote patient monitoring. However, challenges like scalability, standardization, privacy, and security must be addressed carefully as CCNs develop. Other trends, such as AI-driven edge computing, combining with 5G and beyond technologies, explainable AI for administration, predictive maintenance, and evolving into self-learning networks, will define CCNs' future. To overcome obstacles and guide CCNs toward a smart, flexible, and reliable connected future, researchers, company leaders, and regulators must collaborate. In addition to meeting growing communication needs, CCNs may spark innovations that restructure the digital world and usher in an era of intelligent, responsive connectivity as development continues.

## References


1 Iqbal, A., Tham, M.L., Wong, Y.J. et al. (2023). Empowering non-terrestrial networks with artificial intelligence: a survey. *IEEE Access* 11: 100986–101006. https://doi.org/10.1109/ACCESS.2023.3314732.

2 Noman, H.M.F., Hanafi, E., Noordin, K.A. et al. (2023). Machine learning empowered emerging wireless networks in 6G: recent advancements, challenges and future trends. *IEEE Access* 11: 83017–83051. https://doi.org/10.1109/ACCESS.2023.3302250.

3 Bariah, L., Mohjazi, L., Muhaidat, S. et al. (2020). A prospective look: key enabling technologies, applications and open research topics in 6G networks. *IEEE Access* 8: 174792–174820. https://doi.org/10.1109/ACCESS.2020.3019590.

4 Kuruvatti, N.P., Habibi, M.A., Partani, S. et al. (2022). Empowering 6G communication systems with digital twin technology: a comprehensive survey. *IEEE Access* 10: 112158–112186. https://doi.org/10.1109/ACCESS.2022.3215493.

5 Kaneko, M., Ly Dinh, T.H., Kawamura, K. et al. (2023). Wireless multi-Interface connectivity with deep learning-enabled user devices: an energy efficiency perspective. *IEEE Network* 37 (3): 132–139. https://doi.org/10.1109/MNET.123.2100766.

6 Araniti, G., Iera, A., Pizzi, S., and Rinaldi, F. (2022). Toward 6G non-terrestrial networks. *IEEE Network* 36 (1): 113–120.

7 Alhameed, M. and Hossain, M.A. 2023). Rapid detection of pilgrims whereabouts during hajj and umrah by wireless communication framework: an application AI and deep learning," *2023 International Conference on Emerging Smart Computing and Informatics (ESCI)*, Pune, India (01–03 March 2023). IEEE, pp. 1–6, https://doi.org/10.1109/ESCI56872.2023.10099969.

8 Azari, M.M., Solanki, S., Chatzinotas, S. et al. (2022). Evolution of non-terrestrial networks from 5G to 6G: a survey. *IEEE Communications Surveys and Tutorials* 24 (4): 2633–2672.

9 Janjua, M.B., Duranay, A.E., and Arslan, H. (2020). Role of wireless communication in healthcare system to cater disaster situations under 6G vision. *Frontiers in Communications and Networks* 1: 1–10.

**10** Vaezi, M., Azari, A., Khosravirad, S.R. et al. (2022). Cellular wide-area and non-terrestrial IoT: a survey on 5G advances and the road toward 6G. *IEEE Communications Surveys and Tutorials* 24 (2): 1117–1174.

**11** Rahman, I., Razavi, S.M., Liberg, O. et al. (2021). 5G evolution toward 5G advanced: an overview of 3GPP releases 17 and 18. *Ericsson Technology Review* 2021 (14): 2–12.

**12** Kota, S. and Giambene, G. (2021). 6G integrated non-terrestrial networks: emerging technologies and challenges, *2021 IEEE International Conference on Communications Workshops (ICC Workshops)*, Montreal, QC, Canada (14–23 June 2021). IEEE, pp. 1–6.

**13** Giordani, M. and Zorzi, M. (2021). Non-terrestrial networks in the 6G era: challenges and opportunities. *IEEE Network* 35 (2): 244–251.

**14** Giuliano, R. and Innocenti, E. (2023). Machine learning techniques for non-terrestrial networks. *Electronics* 12 (3): 652.

**15** Wang, C.-X., Renzo, M.D., Stanczak, S. et al. (2020). Artificial intelligence enabled wireless networking for 5G and beyond: recent advances and future challenges. *IEEE Wireless Communications* 27 (1): 16–23.

**16** Fadlullah, Z.M., Mao, B., and Kato, N. (2022). Balancing QoS and security in the edge: existing practices challenges and 6G opportunities with machine learning. *IEEE Communications Surveys and Tutorials* 24 (4): 2419–2448.

**17** Song, L., Hu, X., Zhang, G. et al. (2022). Networking systems of AI: on the convergence of computing and communications. *IEEE Internet of Things Journal* 9 (20): 20352–20381.

**18** Mahmood, M.R., Matin, M.A., Sarigiannidis, P., and Goudos, S.K. (2022). A comprehensive review on artificial intelligence/machine learning algorithms for empowering the future IoT toward 6G era. *IEEE Access* 10: 87535–87562.

**19** Letaief, K.B., Chen, W., Shi, Y. et al. (2019). The roadmap to 6G: AI empowered wireless networks. *IEEE Communications Magazine* 57 (8): 84–90.

**20** Yang, Z., Chen, M., Wong, K.-K. et al. (2022). Federated learning for 6G: applications challenges and opportunities. *Engineering* 8: 33–41.

**21** Yang, H., Zhu, Y., and Liu, J. (2019). Energy-constrained compression for deep neural networks via weighted sparse projection and layer input masking. *Proceedings of the International Conference on Learning Representations (ICLR).*

**22** Dinh, T.H.L., Kaneko, M., Wakao, K. et al. (2021). Towards an energy-efficient DQN for user association in Sub6GHz/mmWave integrated networks. *Proceedings of the IEEE Mobile Sensing and Networking Conference*, pp. 1–8.

**23** Zappone, A., Debbah, M., and Altman, Z. (2018). Online energy-efficient power control in wireless networks by deep neural networks. *Proceedings of the IEEE International Workshop on Signal Processing Advances in Wireless Communications*, pp. 1–5.

**24** Sesto-Castilla, D., Garcia-Villegas, E., Lyberopoulos, G., and Theodoropoulou, E. (2019). Use of machine learning for energy efficiency in present and future mobile networks. *Proceedings of the IEEE Wireless Communications and Networking Conference*, pp. 1–6.

**25** Hossain, M.A., Osman, M.H., Hamdan, A.A. et al. (2023). FERLP: facial emotion recognition based on landmark points using artificial intelligence and machine learning. *2023 14th International Conference on Computing Communication and Networking Technologies (ICCCNT)*, Delhi, India (06–08 July 2023). IEEE, pp. 1–6, https://doi.org/10.1109/ICCCNT56998.2023.10308392.

**26** Abdalraheem, M.H.O., Hossain, M.A., Hamdan, A.A. et al. (2023). Estimation of facial emotion based on landmark points by applying artificial intelligence and machine learning. *2023 7th International Conference On Computing, Communication, Control And Automation (ICCUBEA)*, Pune, India (18–19 August 2023). IEEE, pp. 1–6, https://doi.org/10.1109/ICCUBEA58933.2023.10392279.

**27** Eishah, H. et al. (2024). Lightweight and high accuracy facial landmark detection network by applying artificial intelligence. *8th International Conference on Information System Design & Intelligent Applications (ISDIA-2024)*, Dubai (03–04 January 2024). Springer.

**28** Mohd, A.H., Metwally, A., Khan, A. et al. (2024). Cardiovascular disease prediction with convolutional neural networks and hyperparameter tuning. *8th International Conference on Information System Design & Intelligent Applications (ISDIA-2024)*, Dubai (03–04 January 2024). Springer

**29** Hossain, M.A., Amin, S., Eishah, H. et al. (2024). Deep learning for predictive modeling of cardiovascular disease: optimizing neural networks. *8th International Conference on Information System Design & Intelligent Applications (ISDIA-2024)*, Dubai (03–04 January 2024). Springer.

**30** Sharma, M.K., Zappone, A., Assaad, M. et al. (2019). Distributed power control for large energy harvesting networks: a multi-agent deep reinforcement learning approach. *IEEE Transactions on Cognitive Communications and Networking* 5 (4): 1140–1154.

**31** Liu, D., Wang, L., Chen, Y. et al. (2016). User association in 5G networks: a survey and an outlook. *IEEE Communication Surveys and Tutorials* 18 (2): 1018–1044.

**32** Dinh, T.H.L., Kaneko, M., Wakao, K. et al. (2020). Reinforcement learning-aided distributed user-to-access points association in interfering networks. *Proceedings of the IEEE Global Communications Conference*, pp. 1–6.

# 12

# Enhancing Scalability and Performance in Networked Applications Through Smart Computing Resource Allocation

*Araddhana Arvind Deshmukh*[1], *Shailesh Pramod Bendale*[2], *Sheela Hundekari*[3], *Abhijit Chitre*[4], *Kirti Wanjale*[5], *Amol Dhumane*[6], *Garima Chopra*[7], *and Shalli Rani*[7]

[1]*Department of Computer Science & Information Technology (Cyber Security), Symbiosis Skill and Professional University, Kiwale, Pune, Maharashtra, India*
[2]*Department of Computer Engineering, NBN Sinhgad School of Engineering, Pune, Maharashtra, India*
[3]*MITCOM, MCA Department, MIT ADT University, Loni Kalbhor, Pune, Maharashtra, India*
[4]*Department Electronics and Telecommunication Engineering, Vishwakarma Institute of Information Technology, Pune, Maharashtra, India*
[5]*Department Computer Engineering, Vishwakarma Institute of Information Technology, Pune, Maharashtra, India*
[6]*Computer Science and Engineering Department, Symbiosis Institute of Technology, Lavale, Pune, Maharashtra, India*
[7]*Chitkara University Institute of Engineering and Technology, Chitkara University, Rajpura, Punjab, India*

## 12.1 Introduction

Having the ability to expand smoothly and perform optimally is increasingly important in the ever-changing world of connected applications. Both companies and consumers rely on connected applications to help people communicate, work together, and share data as the digital world grows quickly. However, handling and distributing resources gets harder because of the growing variety and complexity of these applications. As a result, the idea of intelligently placing computing power has emerged as a revolutionary approach with the power to hugely change how connected applications are made, expanded, and improved [1].

The basic idea is to make systems aware of themselves and able to improve their performance. These systems can give out resources based on real-time watching of user behavior, app behavior, and what's happening around them. Being able to boost how well things can scale is an important part of smart computing resource allocation, the structure is illustrated in Figure 12.1. Traditional ways to get scaling

*Current and Future Cellular Systems: Technologies, Applications, and Challenges*, First Edition.
Edited by Garima Chopra, Suhaib Ahmed, and Shalli Rani.

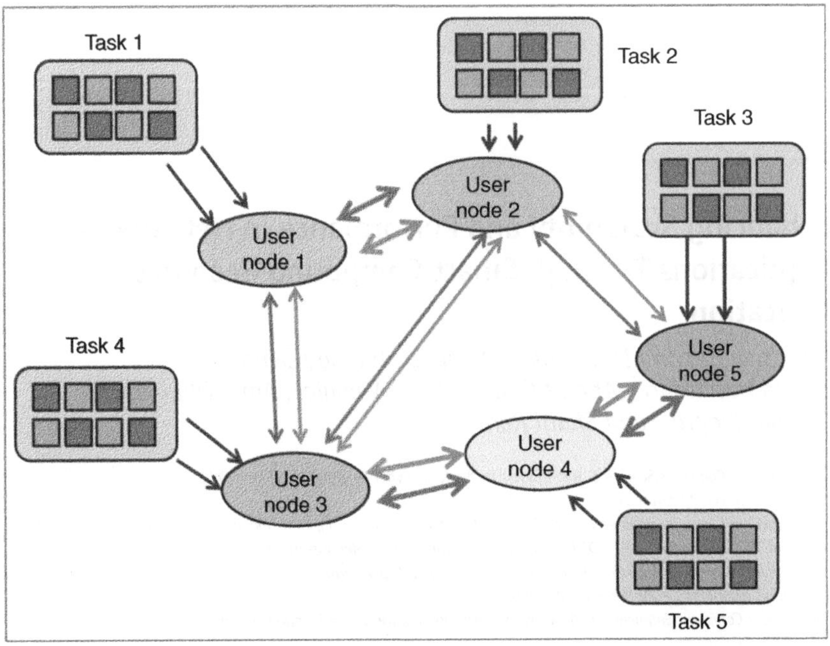


**Figure 12.1** Representation of node and resource allocation structure.

often wastefully use resources when demand is low because resources are given to handle peak needs. However [4], the ability to change resources based on an app's changing needs, known as dynamic scaling, is let by smart computing. Because companies only pay for the resources they really use, this helps lower costs while also using resources the most.

Intelligent systems are improving how apps use resources. They can spot busy and slow areas, share resources better, and fix potential problems by always watching how well things work. Keeping apps fast is important with so much competition online [5]. This helpful monitoring also makes the user experience better. Smart resource allocation does more than just make apps bigger and faster. Being able to quickly change plans when needed saves energy, uses what we have well, and helps the environment. Groups can use less power when demand is low and save money and emissions by adjusting resources based on need. In these changing times, networked apps must be able to grow well and work right. Smart computing resource allocation is an important new way to solve these challenges [2]. This report looks closely at how smart computing works, the tools it uses, and how it could affect how networked systems perform and grow. The goal is to help you understand all the ways smart resource allocation is changing apps and their future as new networks develop.

### 12.1.1 Scope and Objectives

This goal is to carefully look at and think about how smart computing resource giving out is used as an important plan to make networked applications work better and get bigger. The investigation looks at a lot of smart computing things, such as data number work, machine learning, and artificial intelligence (AI), among others, to completely look at how resources can be used in the best way.

- Study the current condition of technologies, including AI programs, machine learning designs, and information examination instruments that help brilliant processing asset circulation, to completely comprehend their qualities and shortcomings. These innovations are utilized to consequently convey assets to PCs, servers, and gear with the end goal of handling occupations and requests from clients and applications productively.
- The ways smart computing resource allocation could be used in different types of connected applications and networks. Some examples are communication programs that need responses quickly, cloud and edge computing systems that share resources, and Internet of Things devices that all work together. Each area offers chances to improve how resources are assigned, but they also have difficulties to deal with because of how they are set up and what people need from them.
- Evaluate how well smart computing resource allocation systems can change quickly in real-time, looking closely at how fast they react to sudden shifts in what programs need and things outside that affect them.
- **Cost Savings:** Examine how putting money toward smart technology affects company finances, focusing on saving money by adjusting what's needed when it's needed, using what we have well, and making sure we get value for what was spent.
- Investigate how networked devices share resources and how this sharing impacts energy consumption. Study how tools that adjust resources based on need can help save power and help the environment. Look at how changing the way systems divide up things such as processing and bandwidth between linked devices can cut energy waste.

### 12.1.2 Objectives

#### 12.1.2.1 Key Goals of This Study

Learn all about how technology shares its abilities with different programs working together online. Know how computers give each job just what it needs to get done. Find out which tools let systems split up tasks and come together to finish big projects.

- Resource allotment in smart computing has the capacity to create organized apps much more productive and fast. Smart allotment makes any doubt that each work gets the computing power control, memory, and speed it needs by consequently allotting assets based on real-time requests. This makes the framework run superior in general.
- Although intelligent resource allocation in computing has benefits, it also comes with some challenges that must be overcome. Smart selection can be hindered by technical issues such as latency, resource contention, and network congestion, and practical issues such as monitoring complex platforms, ensuring different systems work together, and ensuring compliance with security and privacy standards make it more difficult to set up.
- **Cost-Benefit Analysis:** Figuring out investing in smart allocation methods gives a better return on money than investing in traditional static methods by putting a number on these factors. Based on economic factors, this study will help businesses decide if they should adopt smart computer technologies.

The study aims to meet these targets to push forward the ongoing discussion regarding the ground-breaking abilities of clever computing resource allocation within interconnected programs and provide wise guidance to scientists, experts in the business, and people in charge.

## 12.2 Foundations of Smart Computing Resource Allocation

Smart computing resource allocation uses cutting-edge methods to efficiently share computing tools across connected programs. It is done by merging advanced strategies, study methods, and smart solutions. Smart computing applies data investigation, machine learning, and man-made brainpower (AI) to create adjustable and self-aware frameworks that can react rapidly to shifting circumstances [3]. AI, which empowers robots to mimic human reasoning, is fundamental to this structure. AI algorithms are pivotal in resource allocation as they recognize and anticipate usage examples, permitting frameworks to adjust to changes in needs. Machine learning is a core piece of AI as it enables frameworks to enhance execution by adapting and gaining from input and ordeal. The versatile learning capacity, signified by [9], is particularly fundamental in dynamic conditions where the workloads of interconnected applications experience changes.

Analyzing data is key to gathering facts, looking closely at details, and learning new things from large amounts of interconnected information. Advanced study helps systems notice repeating examples, strange events, and share resources well. Checking user interactions and how well things work in real time allows for good

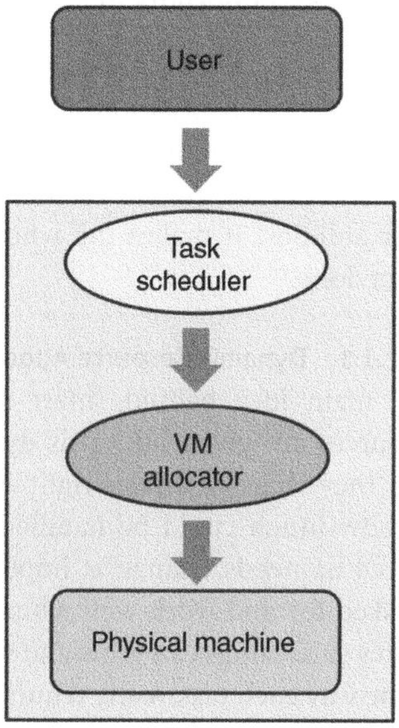


**Figure 12.2** Resource allocation in cloud environment.

changes, as shown in Figure 12.2. This guarantees reacting quickly and doing well, as well as using what you have best. How resources are shared changes a lot. Long ago sharing things the same way often failed when workloads and users changed a lot. Smart computing resource sharing begins a new way of freely changing what you have when needs appear in real time. This makes the most of what you have by wasting less when demands are low and making sure things work well during busy times. It is also important to change when needed. Systems that share what they have well can react fast to sudden shifts in what programs need or unexpected things. Being able to change is key for keeping things working their best and users happy, especially when linked programs face troubles or use changes suddenly.

The integration of security aspects is a key part of smart computing resource distribution [12]. The intelligent resource distribution process considers security standards to ensure the protection of sensitive information and critical processes. Furthermore, using machine learning methods can help in identifying and stopping possible dangers, improving the total security situation of networked applications. Being compatible with the present framework and having the capability to effortlessly work with other systems are further important elements. Smart computing resource distribution answers should own the ability to seamlessly interface with diverse networked environments, accommodating different application architectures and technological frameworks. The benefits of intelligent computing can be experienced in various areas and application fields because of this capability to work together seamlessly.

## 12.2.1 Key Concepts in Resource Allocation

The new system aims to improve how shared devices work together. It uses core rules to help networks adjust to busy and quiet times. Resources are given to what they require the most, keeping everything running smoothly in any sort of task. This strategy focuses on key guidelines that determine how computers share their abilities. It makes the whole setup flexible depending on what jobs need to get done.

### 12.2.1.1 Dynamic Resource Allocation

The main idea behind smart computing resource sharing is adjusting how resources are given out. This dynamic resource allocation lets systems change how they share resources right away, unlike older ways that set resources based on how much could be handled before. Being able to change how things are shared as needs change is important, allowing things to work best when lots is asked for and work well when less is asked for. Dynamic resource allocation makes sure things can grow and work at their best by sharing computer resources in the way each task really requires. This sharing ensures working well no matter the workload.

### 12.2.1.2 Artificial Intelligence (AI) in Resource Management

Computer programs are important for sharing things over networks and changing how things are given out. AI uses machine learning to study past use, guess what will be needed later, and give out things in the best way. AI helps smart systems get better at sharing resources by always learning and changing. This makes sure computing power always fits what the program will need next.

### 12.2.1.3 Using Real-Time Analytics to Track Performance

Real-time analytics is vital for achieving top functions in networked systems. The sharing of resources in smart computing depends on constant watching and studying of environmental factors, how folks act, and performance measures. By using real-time analytics, systems can find clogs, judge the effect of resource giving choices, and actively change settings to keep the top function [16, 17]. This idea makes certain the system can foresee and reduce possible problems before they negatively influence function, while still reacting to the current situation.

### 12.2.1.4 Scalability and Elasticity Measures

The idea of elasticity is important because it shows how well a system can flexibly change its resources as needed. Metrics that measure scalability are important for seeing what a system can do. The goal of smart computing resource allocation is to create and test measures that show how good a system is at handling growing amounts of work without wasting resources when less work needs to be done. Developers make resource allocation algorithms for connected apps better by following these standards. They give a way to numerically see how well scaling is achieved.

#### 12.2.1.5 Mechanisms of Adaptive Learning

The smart allocation of computing resources requires including adjustable studying methods. Machine learning allows programs to gain knowledge through trying things, permitting them to change how they allot assets based on past information and how individuals act. Through adaptive studying, the framework will consistently enhance its capacity to oversee various loads and circumstances, consequently improving its asset allotment capacities after some time.

#### 12.2.1.6 Security-Driven Resource Allocation

Keeping things safe is very important when giving out things people need. The idea of giving out things in a safe way means thinking about protection when deciding who gets what. Smart sharing of resources makes important work and information secure by remembering to protect things when sharing resources. Machine learning can help spot dangers and make networked programs stronger.

### 12.2.2 The Evolution of Scalability and Performance in Networked Applications

Technology is changing how apps work on networks. Networks now need to change a lot to keep up with more people using more things online. Apps used to stay the same but now they have to get better at using computer power in flexible ways. This helps apps get faster and bigger without problems. It shows how companies found new ways to share resources that adjust to apps needing more or less at different times. These ideas helped create smart computing, which is better than the old ways apps used to work.

#### 12.2.2.1 Conventional Static Resource Allocation

Some early networks set up resources the same way each time. They gave systems more than needed sometimes and not enough other times because the amount was based on how busy things might get at their busiest. This method couldn't change with changing demands. Networks now can adjust what they give systems depending on how much work needs to be done at any time.

#### 12.2.2.2 The Arise of Scalability Issues

Some challenges arose as more programs joined networks and had different needs. When information and user activities grew very quickly, old plans struggled to keep up. This led to slowdowns and bad user experiences when lots of people used them at once. It was obvious new ways were needed to share resources as needed at the moment.

#### 12.2.2.3 The Cloud Paradigm and Dynamic Resource Allocation

Cloud computing allows computers to adjust based on need. Earlier, computers had set amounts of memory and processing power. This caused issues when more was needed. Clouds let resources expand or shrink depending on how they are being used. This helped solve problems with fixed amounts that couldn't change. It made computers more flexible. They could adjust up or down automatically instead of being stuck at one level. This helped address limitations with the old way of assigning resources.

#### 12.2.2.4 Using Smart Computing to Allocate Intelligent Resources

Smart devices are changing how we use technology for better results and growth. Ideas such as artificial thinking, machine learning, and real-time data reviews show new ways to share tools. Systems can now plan ahead for what's needed, understand past facts, and adjust how they assign what's required thanks to machine learning steps. Real-time data reviews help us make choices before problems by teaching us how people use things, how apps work, and what's around us.

#### 12.2.2.5 Real-Time Adaptation and Predictive Scaling

The new smart computing resource allocation uses real-time changes and future predictions to adjust resources. Unlike waiting until after problems happen, it looks at past data to see what may come next. This helps prepare resources before users need them. The system can make resources bigger or smaller before demand rises and falls. That means the system works well and users stay happier, even during very busy times.

#### 12.2.2.6 Scalability Beyond Traditionally Assigned Limitations

Applications have grown in new ways due to advances in technology. Programs can now use outside sources and spread over different areas, instead of just big computer centers. Whether adding more to what's there or spreading tasks between locations, apps can easily change to work in different situations thanks to smart computing sharing resources where needed. This enables applications to perform optimally regardless of their setup.

#### 12.2.2.7 Automation and Autonomy's Role

Some choices and tasks are now made by computers and machines. Decision-making steps are done by computer programs that can share resources in better ways without people needing to help. Systems that can work on their own are able to give out what is needed at the right time in a smart way. They make everything adjust to changes without people controlling everything.

## 12.3 Dynamic Resource Allocation Strategies

The most important parts of the smart computing resource allocation system are the rules for sharing resources, which are key to boosting how the connected systems work. These tactics use computer smarts, learning from examples, and live data to make good choices about spreading computing power around. Networks can change to fit changing needs, use resources the best, and fix issues before they happen because the tactics can change. Here, we look at flexible ways to share resources and how they impact how well things run and how large systems can grow. Table 12.1 summarizes the allocated resources, specific

**Table 12.1** Summary of resources allotted, technological situations, problems or characteristics taken into account, and the security fixes suggested in the relevant research.

| Resources technology scenario | Issues/attributes consideration | Security solution |
|---|---|---|
| Computation, radio resources | 5G MEC distributed | Data breach, security layer |
| Data rate, virtual resources | 5G HetNets distributed | Data breach, constraint |
| Data | 5G distributed | Data breach, constraint |
| Data | 5G distributed | Data breach, constraint |
| Network slices | SDN/NFV-enabled 5G distributed | DDoS attack, constraint |
| Bandwidth | IoT distributed | DDoS attack, constraint |
| Computation | Cloud computing distributed | DDoS attack, constraint |
| Network slices | 5G RAN centralized | Side channel attack, constraint |
| Spectral resource | D2D communications distributed | Eavesdropping, constraint |
| Homogenous and heterogeneous clusters | Cluster computing parallel | CIA triad, constraint |
| Radio resource | Cognitive radio networks distributed | Authentication, constraint |

technical scenarios, challenges considered, and recommended security solutions from related research, providing a comprehensive overview of the intelligent allocation of computing resources and its effectiveness in different situations.

Predictive scaling uses past data and patterns to foresee future needs. This approach uses machine learning to study and understand changes in needs over time. By expecting upcoming increases in demand, the system gets ready for busy times or changes in how people use things. Predictive scaling improves how responsive programs are and saves money by only adding more power when needed, not spending when times are slow. The automatic control of computer strength is meant to react to real needs, letting connected programs flexibly change their size within set rules or boundaries, the allocation of resources overview is illustrated in Figure 12.3.

Some parts of computer programs change size depending on how much work needs to be done. This is called auto-scaling. It lets programs adjust themselves without people fixing them. When more or less work needs to get done, auto-scaling gives or takes away computing power. This helps keep things running well no matter how busy or quiet it is. It keeps everything balanced so programs can do their best no matter what's going on.

Sharing work between devices helps websites. Traffic coming to a site is split between different computers. No single computer gets too busy. This makes sure computers stay fast. The site also uses all its tools well. Traffic splitting changes as needed. Measurements track how busy each computer is. Traffic is sent where it can be helped the quickest.

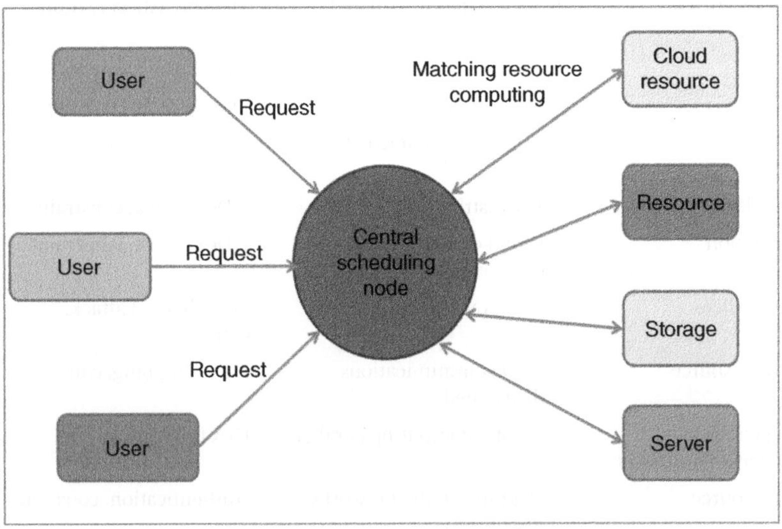


**Figure 12.3** Overview of allocating resources and scheduling tasks.

Dynamic provisioning lets resources like virtual machines and containers change as needed. It makes sure apps get what they need when work is high or low. This helps apps work well with different amounts of work. It's very helpful in cloud computing. There the resources can change fast based on what apps need in the cloud.

## 12.3.1 Static vs. Dynamic Resource Allocation

### 12.3.1.1 Static Resource Allocation

Distributing computer equipment in advance aims to match needs and availability. Leaders allot machines according to past usage peaks and troughs. This method remains steady no matter what user volume changes. While simple to oversee, it can squander supplies. Allotments may exceed requirements when the activity is quiet. Yet shortfalls could develop when workloads surge. Using tools effectively across fluctuating requirements challenges organizers to think flexibly.

### 12.3.1.2 Dynamic Resource Allocation

On the other hand, sharing computer power flexibly is known for its ability to quickly change resources according to what needs to be done. This way, networks can freely give more or less resources depending on how busy or quiet things are. They make the best use of what's available. The heart of Smart computing is sharing power freely, which helps connected programs easily handle different jobs. It keeps things running well and lets them deal with unexpected changes.

## 12.3.2 Adaptive Resource Allocation Algorithms

The goal of adaptive resource allocation algorithms is to dynamically change how resources are shared based on current situations. These algorithms are extremely important for smart computing resource allocation because they enhance the system's ability to increase how much it can do and make it work better. Load balancing algorithms are meant to stop workloads from being uneven by spreading new visitors among many servers or things. Adaptive load balancing algorithms constantly adjust the spreading depending on real-time facts, guaranteeing the best application performance and effective use of resources. Based on anticipating analytics, these algorithms guess future needs. These algorithms can actively grow resources before demand jumps a lot, increasing quickness and decreasing wasted resources by looking at past information and patterns.

AI programs that learn to manage resources through experience are called adaptive algorithms. By studying old information and how people act, these programs become smarter over time at sharing things. They adjust the way they give out resources as situations change, refining their methods to work better in response.

Through this continuous adaptation and improvement, the algorithms enhance their capabilities to maximize benefits.

### 12.3.3 Machine Learning Approaches in Resource Allocation

Machine learning approaches use smarts to help pick how to share things. They use steps to share things in the best way and learn from what they know. If they learn new stuff, they can change how they share things. Here are some important ways machine learning shares things:

- Reinforcement learning teaches computers to make choices step-by-step. It rewards or punishes choices based on results. This helps algorithms learn the best order for decisions. Reinforcement learning can adjust how it uses things, considering what happened before. It aims to use resources in the most effective way.
- **Supervised Learning:** Algorithms are taught using examples to make predictions. They use historically labeled data to be trained. Supervised learning can anticipate needs by examining past usage. It helps prepare for growth by predicting future needs based on how much was used before.
- **Learning Without Guidance:** Unsupervised learning examines data without examples to follow. It seeks patterns and connections in the information. This type of learning can find how resources are used. It detects unusual or irregular resource activity that may signal security concerns or issues with performance. It assigns resources to where they are most needed.
- **Deep Learning:** Deep learning uses networks that learn from complex examples. It models long chains of causes and effects using computer systems with many layers, such as neurons in a brain. These "deep neural networks" can find solutions to hard problems such as sharing limited things between many parts of a busy connected system. They help understand how everything interacts in detail, even when the system acts in confusing ways under pressure.

Our research into how computers give out memory and processing time shows how important it is for technology to meet the needs of modern internet programs. The basis for smart computing resource giving is built from rules that change how resources are shared and ways of learning from data, which help computers, grow well as needs change, do their best, and adjust smoothly when things change.

## 12.4 Intelligent Load Balancing Techniques

It is important for networks to have smart methods of moving traffic among servers. These techniques help make good use of resources and help apps work better. Adaptive load balancing watches things such as how the network is doing, how busy the servers are, and how quickly servers give answers. It uses this data to adjust how new traffic is spread when things change. Load balancers can send traffic to less crowded servers if a server gets slower or too busy. This helps prevent problems and keeps work divided evenly. Moreover, machine learning makes load balancing more complex. These methods can predict and choose well because they learn from what happened before, how people use things, and how the system did its work. Balancers can change their plans based on past results using reinforcement learning, which is a type of machine learning. This helps them get better at sharing traffic and resources over time.

### 12.4.1 Load Balancing in Networked Environments

Load balancing spreads incoming network traffic across multiple computers or servers. Its goal is to use computer resources better and make networked apps work better. To stop slowdowns and keep things quick, load balancing shares workloads evenly between servers. This keeps any one server from slowing others down. Load balancing in networks aims to hand out work equally, as shown in Figure 12.4.

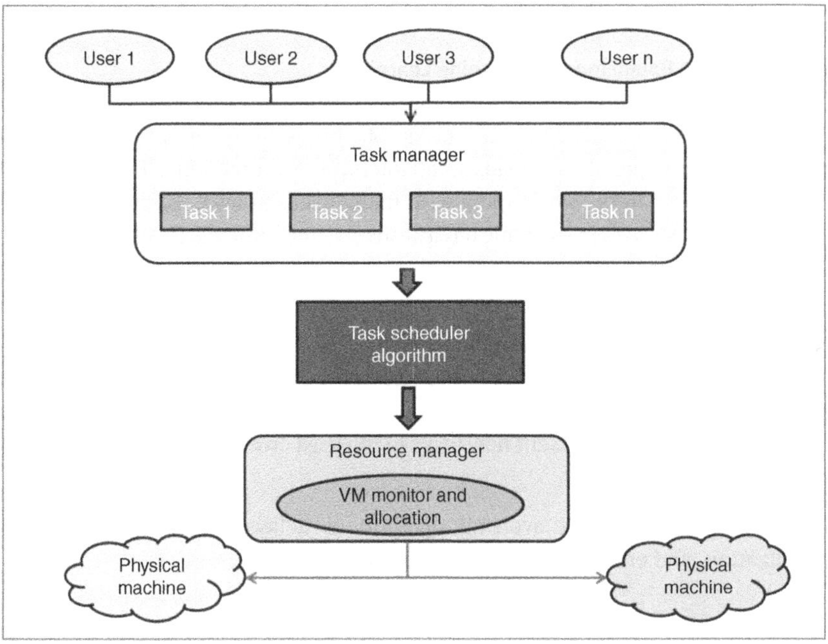


**Figure 12.4** Representation of load balancing in networked environments.

The goal is to evenly distribute work among servers. It aims to prevent any single server from becoming overloaded with too many tasks. This spreading of work is particularly useful when multiple servers collaborate on running an application or delivering a service. Load balancing avoids burdening servers with more than they can handle. It also ensures that each server contributes its appropriate portion to meet user demands.

## 12.4.2 Importance of Load Balancing in Scalability

Load balancing helps make sure computer systems, networks, and apps work well when busy. It splits work between servers so one doesn't get too full. This lets everything run fast no matter how many people are using it. Load balancing splits tasks so servers don't get crammed with too much to do. It spreads work around so systems can handle lots of people without slowing down.

Spreading work evenly stops any single part from having too much to handle. Load balancing shares tasks so each part gets its fair share. This means no part is asked to do more than it can. If work is spread out, nothing gets too busy or too full. The whole system runs better when no piece is too stressed. Parts work at a steady pace when we divide work up sensibly. When no piece is over its head, the whole thing hums along nicely.

### 12.4.2.1 Load Balancing with Machine Learning

Load-balancing methods use past records and current information to change how computer power is shared between tasks. Machines studying old and new data can learn to best spread workloads so no job has to wait too long. This sharing of work keeps computers busy helping with many needs at once. Embrace continuous learning, optimize in dynamic environments, and embrace flexible workloads for maximum benefits.

### 12.4.2.2 Adaptive Load Balancing Algorithms

The computer programs that direct internet traffic changed how jobs were shared between servers and networks. They watched numbers that showed how busy everything was. Then they sent new tasks to the least crowded places, as described in Table 12.2.

- Advantages encompass averting operational delays, maximizing resource utilization, and ensuring instantaneous responsiveness.

**Table 12.2** Comparison of smart load balancing algorithms.

| Algorithm | Key parameters for comparison |
|---|---|
| Predictive load balancing | Prediction accuracy, adaptability to varying workload trends, proactive resource allocation, and impact on response times. |
| Machine learning-based load balancing | Learning speed, adaptability, accuracy in decision-making, resource optimization, and overhead in training models. |
| Adaptive load balancing | Real-time responsiveness, effectiveness in preventing bottlenecks, adaptability to sudden changes, and resource distribution efficiency. |

## 12.5 Real-Time Monitoring and Feedback Mechanisms

The distribution of computer assets relies greatly on observing and reactions in actual time, which improves adaptability, productivity, and general execution of joined applications. A few investigations into watching immediately give an understanding of proactive observing for asset designation, input circles, and dynamic basic leadership.

### 12.5.1 Proactive Monitoring for Allocation of Resources

Careful watching involves keeping an eye on many things all the time, such as how busy the servers are, how well the apps work, and the network setup. Making good choices about sharing resources and guessing what will happen later relies a lot on this information. Systems that carefully watch are better ready to see changes coming, find places that could slow down, and adjust the resource plan before things get worse. Predictive algorithms can use data on site usage patterns to allocate resources in anticipation of expected busy periods.

### 12.5.2 Decision-Making and Feedback Loops

Making choices and checking results needs feedback. Loops help with this. They gather details about what is happening now. People look at the details. Then they change how things are shared because of what they saw. Loops often use machine learning. It learns from past choices. It adjusts to how things change.

The system will get better at sharing things in a smart way over time. This happens because of the feedback that goes back and forth.

### 12.5.3 Real-Time Monitoring

**Case Study 12.1**

**Netflix**

Netflix carefully watches how people use its streaming service. To make sure videos play smoothly, Netflix pays close attention to what viewers watch and how fast their internet is. It can lower or raise the quality of videos and adjust resources as needed. Constant checking helps prevent videos from pausing to load and makes the best use of computers. This active watching helps Netflix provide a streaming service that just works.

**Case Study 12.2**

**Uber**

Uber's watchful system observes its complex ride system closely. It constantly follows how many people need rides, traffic, and where drivers and people in cars are. Uber can change which drivers take people places, pick better roads, and make the whole thing work better because it gets updates all the time.

Large cloud networks such as Amazon Web Services (AWS) and Azure rely on constant tracking to manage their big setups. To watch how people use systems, how long networks take, and how servers work, they use advanced watch programs. These networks can flexibly change what they have based on need thanks to the current data that runs smart computer distribution rules. How well smart computing resource allocation works depends on tracking what happens now and sharing what they learn. Changes in how busy systems are may be expected before they happen from watching ahead, and sharing what they notice makes sure the system always learns and changes to fit. Stories from cloud networks, Netflix, and Uber show how these ideas can be used in real life to better share tools and make networked programs do better altogether.

## 12.6 Case Studies and Best Practices

### 12.6.1 Cloud-Based Resource Allocation

**Case Study**

**Amazon Web Services (AWS)**

To suit the different needs of its users, AWS uses smart cloud-based ways to share resources. Auto scaling is an AWS tool that automatically changes how many computers are used based on user rules. AWS predicts when more people will visit and adds computers before they are needed. It also watches what is happening and adjusts resources all through its huge cloud setup.

Best Practices:

- **Cost Optimization**: Putting procedures in place ensure that resources are used efficiently and to minimize costs during times of low demand.

### 12.6.2 Edge Computing and Resource Optimization

**Case Study**

**Microsoft Azure IoT Edge**

Microsoft Azure IoT Edge allows devices on the network's outskirts to better organize their resources. This program lets machines close by performing quick local data analysis themselves. It loads AI designs directly into internet-connected appliances. This cuts down waiting time, uses edge equipment more efficiently, and decreases how much information must be sent to faraway servers in large clusters.

Best Practices:

- **Distributed Computing:** Lessen reliance on centralized cloud computing by allowing processing to occur across many edge devices.
- **Adaptive Learning:** Utilize intelligent programs that study regional details and alter approaches to evolving scenarios.

### 12.6.3 High-Performance Computing (HPC) Environments

**Case Study**

**NASA's Pleiades Supercomputer**

NASA's Pleiades supercomputer efficiently shares its power with many science tasks. It uses management strategies to schedule workloads and jobs. This ensures computing power goes where projects need it most based on importance and need. The supercomputer constantly watches itself to make sure resources stay busy helping the most important research. If a task needs more resources or another task has extra, it adjusts to use all of its abilities for advancing human knowledge.

Best Practices

Parallel Processing: Workloads can be spread across multiple computers in an organized way by carefully assigning tasks to resources to best use parallel processing.

The importance of adjusted plans for sharing assets in different processing conditions is appeared by these case thinks about and best strategies. Productive sharing of assets in cloud, fringe, and top execution processing situations requires a blend of versatile learning procedures, decentralized handling, anticipated investigation, and flexible extendibility. These strategies guarantee ideal execution, viable use of PC assets, and smooth extendibility of applications to meet evolving requests.

## 12.7 Security and Privacy Considerations

### 12.7.1 Ensuring Security in Resource Allocation

#### 12.7.1.1 Overview of Security

For sharing limited tools, keeping things safe is important, especially with smart tech places. Because how things are split changes, tight protection steps help stop hacking, unauthorized entry, and facts leaks.

Implement tight access restrictions to ensure that choices about resource distribution can only be impacted or adjusted by approved individuals.

- To protect information as it travels and when saved, use encryption covering the whole path. This shields

- Monitoring tools watch out for strange behavior and find clues of possible security problems or tries to damage how resources are shared. Intrusion detection mechanisms observe networks to spot abnormal actions and see if anyone is trying to access systems without permission or slowly over.

Choosing protected ways for parts of the system to talk will help keep resource sharing safe. Some connections need extra security, while others require less. It's important to pick techniques matching.

## 12.7.2 Privacy Issues with Wise Resource Distribution

### 12.7.2.1 Overview of Privacy

Different kinds of data are gathered and studied to power new technologies. It's important people understand private details are kept private and safe. Companies must be careful with information.

- Gather only required information for allocating resources and store as little private data as possible. Keep collected details to a minimum.
- Use tactics such as assigning fake names and removing identifiable details to shield personal data and decrease the chances someone could connect information to a specific person. Strategies for protecting
- **Transparent Policies:** Inform consumers in a clear and understandable manner about privacy policies, including how their data will be used to allocate resources.

## 12.7.3 Balancing Security and Performance

### 12.7.3.1 Understanding the Art of Balancing Responsibilities

It can be hard to share resources between protection and speed the right way. If safety is skipped, things could be in danger. But focusing too much on safety could cause issues with how well things work.

- **Implement Precise Oversight:** Develop mechanisms allowing managers to finely tune protection based on how crucial budget choices are and how sensitive information is.
- To determine how successfully safety rules allow sharing of things, watch closely how monitoring tools show use over time.
- Implement changing protections for adaptive safety that can automatically change to match the needs of current projects and the latest risks from threats.

Some parts of smart computing plans focus more on protecting information while others aim to use resources well. For programs to go as planned, how well they work and keep people safe must be balanced right. Groups can make sure resources are given out in a way that reacts fast and uses little but also watches for danger and keeps user information private by adding good protections, fixing problems with privacy, and using different methods that adjust based on need. Using all these things together is key so people feel fine and sure about sharing more information online as tech gets smarter and connects everyone in new ways.

## 12.8 Future Trends and Emerging Technologies

### 12.8.1 Resource Allocation and Edge AI

#### 12.8.1.1 Understanding the Basics of Edge AI

A new idea that is altering how resources are shared is edge AI, or placing AI skills directly onto devices. Edge AI diminishes waiting time and how much of the network is used by examining data right on gadgets instead of only relying on centralized cloud systems.

- **Impact of Resource Allocation**:
- **Distributed Processing:** Edge AI brings about a shift toward distributed computing, impacting how resources are assigned. Machine learning models are spread out across many devices at the edges of the network, prompting algorithms to optimize nearby handling of information rather than depending mainly on servers in data centers.
- **Resourceful Distribution:** In Edge AI, allocating tasks involves sharing computational jobs between devices in a productive way to lessen energy use and better general functioning.
- **Adaptive Learning:** Machine learning models located directly where the data is collected can study trends and patterns in the information.

**Future Consequences**

- Resource allocation methods may become more adaptable and mindful of certain situations as edge AI advances. Clever tools and gadgets around us will disperse computing power in a way that responds better to what's happening nearby. This will allow devices to adjust how they use things such as processing and memory to match what they need depending on what's going on.

### 12.8.2 Implications for Quantum Computing

#### 12.8.2.1 A Comprehensive Look at the World of Quantum Computing

Quantum computing uses ideas from quantum physics to solve complex problems very fast. It works differently than regular computers. Quantum computers can solve some types of problems much quicker than normal computers.

**The Significance of Allocating Resources**

Quantum superposition allows us to look at multiple options at once, which is why quantum computers are so good at doing lots of work together at the same time. We'll need different plans for sharing things to use all that teamwork.

Sometimes particles connect in ways that change how they work together. This is called quantum entanglement. It affects how parts of quantum computers, called qubits, share resources. Entanglement impacts how qubits distribute processing tasks or data.

- **Quantum Computing Environments:** Designing and improving quantum computing systems may require different plans for distributing resources to quantum programs compared to classical computer programs. Quantum algorithms may function best with resources handled in special ways.

**Future Consequences**

- As quantum computing progresses, the methods used to assign resources will need to adapt to make the most of quantum phenomena. The way computing power is typically shared may have to be reimagined to completely benefit from the unusual link between particles and the immense simultaneous potential offered by quantum technology.

## 12.8.3 Allocating Resources with Blockchain

### 12.8.3.1 Overview of Blockchain

Blockchain can record details about all kinds of information in a way that makes it very difficult or impossible to change, hack, or cheat the system. It doesn't just help with money such as Bitcoin, but can be used for sharing or keeping track of many different things, such as who owns what or gets what part of something.

The Influence of Resource Distribution:

- **Smart Contracts:** Based on predetermined standards, allotment choices for assets can be carried out automatically by using intelligent contracts, which are self-activating agreements containing encoded conditions. This impacts the clear and reliable circulation of possessions.
- Decentralized autonomous organizations,(DAOs) permit teams to jointly decide how to distribute assets without a head in charge. Blockchain technology enables this new way of collaborating through networks. Members in these groups use consensus decision-makingto decide how resources are shared.

**Future Consequences**

As smart contracts and decentralized governance systems become more common, blockchain is likely to have a bigger effect on how resources are shared.

This could lead to processes for handing out resources that are safer, clearer, and not controlled by any one group in many companies. New techs such as Edge AI, quantum computing, and blockchain will greatly change how shares are given out later on. These changes will sway how pieces, optimization, and use of computer shares are managed in decentralized and more complex computer systems. Tactics for portioning out shares will need to evolve as these techs get better so we can use what makes each one unique and address new chances and troubles.

## 12.9 Conclusion

New techniques for distributing limited resources, guided by smart technology, are key to building a digital world that can adapt to challenges. These new technologies will change how computer networks work. Blockchain, AI, and quantum computers allow networks to adjust quickly to changing needs. Constant watching, guessing what will happen next, and learning from experience are key to keeping networks running well, changing easily, and not wasting resources. Examples such as Microsoft Azure IoT Edge and Amazon AWS show good ways to do this. Different sentences help describe adapting to changing times. New ideas such as quantum computing and smart devices with AI can help or cause problems. We must rethink how we share things. Keeping data safe, private, and useful is very important now that information guides what we choose. In today's world, protecting information, respecting privacy, and using resources carefully is very important as data guides more of what we do. By sharing technology duties, keeping track of what's happening now, and adjusting how we use things, we can make systems work together smoothly. Experts believe tomorrow's tech will be linked together tightly but stay safe and work well, sharing what they need to keep improving quickly, changing when needed, and responding to what networks require.

## References


1 Bruschi, R., Davoli, F., Lago, P., and Pajo, J.F. (2019). A multi-clustering approach to scale distributed tenant networks for mobile edge computing. *IEEE Journal on Selected Areas in Communications* 37 (3): 499–514. https://doi.org/10.1109/JSAC.2019.2894236.

2 Aung, N., Dhelim, S., Chen, L. et al. (2023). VeSoNet: traffic-aware content caching for vehicular social networks using deep reinforcement learning. *IEEE Transactions on Intelligent Transportation Systems* 24 (8): 8638–8649.

3 Kang, J.J., Yang, W., Dermody, G. et al. (2020). No soldiers left behind: an IoT-based low-power military mobile health system design. *IEEE Access* 8: 201498–201515.

4 Okwuibe, J., Haavisto, J., Kovacevic, I. et al. (2021). SDN-enabled resource orchestration for industrial IoT in collaborative edge-cloud networks. *IEEE Access* 9: 115839–115854.

5 Yang, Y., Miao, Y., Ying, Z. et al. (2022). Privacy-preserving threshold spatial keyword search in cloud-assisted IIoT. *IEEE Internet of Things Journal* 9 (18): 16990–17001.

6 Qi, S., Wei, W., Cheng, J. et al. (2022). Secure and efficient item traceability for cloud-aided IIoT. *ACM Transactions on Sensor Networks* 18 (4): 1–24.

7 Fu, F., Zhang, Z., Yu, F.R., and Yan, Q. (2020). An actor-critic reinforcement learning-based resource management in mobile edge computing systems. *International Journal of Machine Learning and Cybernetics* 11 (8): 1875–1889.

8 Bebortta, S., Senapati, D., Panigrahi, C.R., and Pati, B. (2022). Adaptive performance modeling framework for QoS-aware offloading in MEC-based IIoT systems. *IEEE Internet of Things Journal* 9 (12): 10162–10171.

9 Naouri, A., Wu, H., Nouri, N.A. et al. (2021). A novel framework for mobile-edge computing by optimizing task offloading. *IEEE Internet of Things Journal* 8 (16): 13065–13076.

10 Jin, Z., Zhang, C., Jin, Y. et al. (2022). A resource allocation scheme for joint optimizing energy consumption and delay in collaborative edge computing-based industrial IoT. *IEEE Transactions on Industrial Informatics* 18 (9): 6236–6243.

11 Yannuzzi, M., van Lingen, F., Jain, A. et al. (2017). A new era for cities with fog computing. *IEEE Internet Computing* 21 (2): 54–67.

12 Sabella, D., Vaillant, A., Kuure, P. et al. (2016). Mobile-edge computing architecture: the role of MEC in the internet of things. *IEEE Consumer Electronics Magazine* 5 (4): 84–91.

13 Taibi, D., Lenarduzzi, V., and Pahl, C. (2017). Processes motivations and issues for migrating to microservices architectures: an empirical investigation. *IEEE Cloud Computing* 4 (5): 22–32.

14 Bhamare, D., Jain, R., Samaka, M., and Erbad, A. (2016). A survey on service function chaining. *Journal of Network and Computer Applications* 75: 138–155.

15 Nitti, M., Pilloni, V., Colistra, G., and Atzori, L. (2015). The virtual object as a major element of the internet of things: a survey. *IEEE Communications Surveys and Tutorials* 18 (2): 1228–1240.

16 Nunes, B.A.A., Mendonca, M., Nguyen, X.-N. et al. (2014). A survey of software-defined networking: past present and future of programmable networks. *IEEE Communications Surveys and Tutorials* 16 (3): 1617–1634.

**17** Clark, C., Li, Y., Zhang, Y., and Jin, D. (2005). Live migration of virtual machines. *Proceedings of the 2nd conference on Symposium on Networked Systems Design & Implementation-Volume 2*, pp. 273–286.

**18** Mahalingam, M., Dutt, D., Duda, K. et al. (2014). Virtual eXtensible local area network (VXLAN): a framework for overlaying virtualized layer 2 networks over layer 3 networks. Aug. 2014, [online] Available: https://tools.ietf.org/html/rfc7348.

**19** Niranjan Mysore, R., Pamboris, A., Farrington, N. et al. (2009). PortLand: a scalable fault-tolerant layer 2 data center network fabric. *ACM Sigcomm Computer Communication Review* 39 (4): 39–50.

**20** Khobragade, P. and Malik, L.G. (2014). A review on data generation for digital forensic investigation using datamining. *IJCAT International Journal of Computing and Technology* 1 (3).

**21** Bruschi, R., Davoli, F., Lago, P. and Pajo, J.F. (2017). A scalable SDN slicing scheme for multi-domain fog/cloud services. *Proceedings of the IEEE Conference on Network Softwarization (NetSoft)*, pp. 1–6.

**22** Schwabe, A. and Holger, K. (2014). Using MAC addresses as efficient routing labels in data centers. *Proceedings of the 3rd Workshop on Hot topics in Software Defined Networking*, pp. 115–120.

**23** Hari, A., Lakshman, T.V., and Wilfong, G. (2015). Path switching: reduced-state flow handling in SDN using path information. *Proceedings of the 11th ACM Conference on Emerging Networking Experiments and Technologies (CoNEXT)*, pp. 36:1–36:7.

# 13

# 5G-Enabled Fusion: Navigating the Future Landscape of Cloud Computing, Internet of Things, and Recommender Systems

*Sheetal Sharma*

*IT- Department, Goswami Ganesh Dutta Santana Dharma College, Chandigarh, India*

## 13.1 Basics of Cloud Computing

Cloud computing is the ability to access computing services available over the internet, including databases, storage, servers, software, etc., as and when required, on a pay as you go basis. According to a definition provided by U.S. NIST (National Institute of Standards and Technology) [1]. Cloud computing is a theory for providing need-based network access into a shared puddle of configurable computing assets, like databases, applications, services, and networks, that may be quickly supplied and released with no administration effort or service provider contact. In the era of 5G connectivity, cloud computing has evolved to offer seamless and high-speed network access. Instead of owning and managing resources like files and computing infrastructure, such as data centers and servers, locally, one can now leverage dynamically scalable and virtualized resources as a service over the Internet via a range of devices, including desktops, smartphones, laptops, and tablets, etc. (Figure 13.1). Cloud computing, in the context of 5G, not only facilitates faster and more efficient access but also supports the integration of diverse devices and applications. Cloud computing adheres to two fundamental latent features: Abstraction and Virtualization. The information regarding the physical aspects, such as location, administration, and implementation of the system being used for or as the service, is abstracted from the cloud user. Distributed and shared resources are virtualized as a system, where resources can be scaled and made accessible from a centralized infrastructure when needed [2].

A) ***Characteristics of Cloud Computing*** [1–4]

**Cost Efficient:** Users can avoid capital expenses of buying software and hardware, setting up and maintenance of local data centers, power consumption

*Current and Future Cellular Systems: Technologies, Applications, and Challenges*, First Edition.
Edited by Garima Chopra, Suhaib Ahmed, and Shalli Rani.

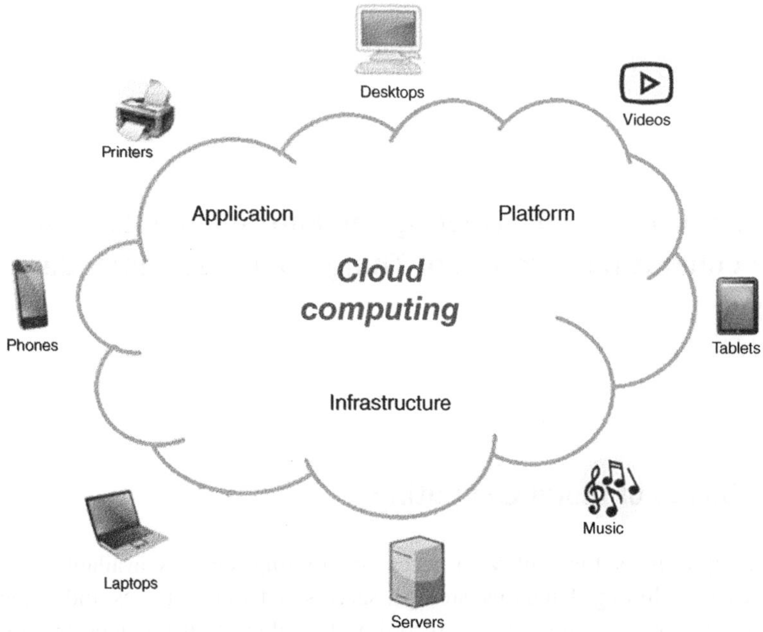


**Figure 13.1** Cloud computing in varied devices [2]/With permission of John Wiley & Sons.

by the servers, and resource managing experts. As the third party holds the cost of implementing and maintaining the infrastructure being used. But if the requirement is predictable, regular and for the long term, in-house setup proves to be more economical.

**Global Access:** Cloud computing elasticities device and geographical independence, users can access the service via the internet from anywhere anytime and through their respective devices (PC, Laptops, or mobile). Internet outrage, can interrupt access to the cloud service and thus affect the work.

**Performance:** IT experts from the service provider make sure the efficient and upgraded condition of the computing hardware and software.

**Security:** Centralization can enhance security, also many Service providers offer far-reaching guidelines, policies, and technology to ensure data security from possible threats. Sensitive data handled by a service that is also used by other competitive organizations can sometimes be a threat.

**Scalability:** Capability to scale up and down elastically any service like storage, bandwidth, or computing power when demand increases or decreases is a significant benefit of cloud computing.

**Productivity:** Use of service concurrently by multiple users and time saved due to automatic access to data and services increase productivity.

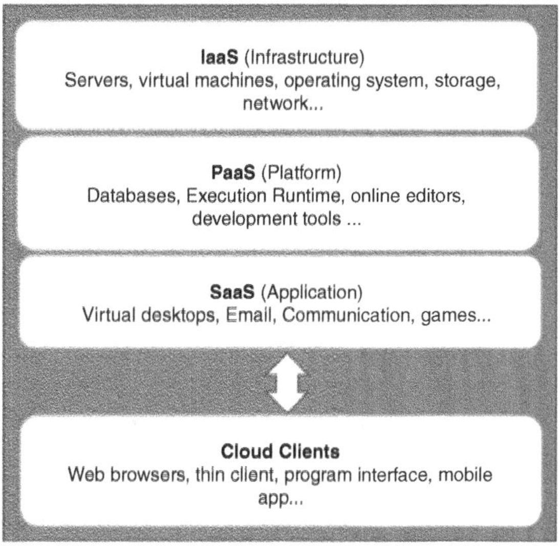


**Figure 13.2** Cloud computing stack [5]/With permission of McGraw Hill Education, Inc.

B) ***Service Models***

Services proposed by cloud computing service providers can be classified into three standard service models as stated by NIST [1, 2] Infrastructure as a Service (IaaS), Platform as a Service (PaaS), and Software as a Service (SaaS). The level of abstraction increases gradually with each model. As they arelayered on top of one another these models are often known as cloud computing stacks [2–4] Figure 13.2.

**Infrastructure as a Service (IaaS):** It is the most fundamental type of cloud computing service. The user rents physical infrastructure as computing resources, operating systems, storage, servers, and networking. The consumer in *IaaS* can only manage the storage, OS, and deployed applications and is restricted from managing cloud infrastructure. e.g., Amazon EC2 [5].

**Platform as a Service (PaaS):** It is the next level in the stack. Besides control over the physical infrastructure, it also facilitates an environment and tools for developers to quickly build and deploy different applications hosted by the cloud provider. e.g., Right Scale and Google [2, 5].

**Software as a Service (SaaS):** A simple cloud that provides software to use hosted by the cloud servers. No control over the infrastructure. Upgradation and maintenance are managed by cloud service providers. Consumers can access the application via the internet using web browsers or a program user interface. e.g., Salesforce and Google App Engine.

C) ***Types of Cloud Computing***

Different types of users demand different types of services; thus, one type of cloud computing cannot serve the purpose. There are three basic types of cloud computing in which a cloud can be deployed: a public cloud, private cloud, and hybrid cloud [1–4].

**Public Cloud**: Services provided by the cloud are publicly accessible over the internet but owned, handled, and maintained by the third party may be private (business or academic) or government organizations or maybe by both. Also, known as the External cloud.

**Private Cloud**: Referred also as an Internal Cloud, enables services explicitly for use of any particular client with multiple consumers with control over infrastructure and data restricted to the organization itself or by the third party.

**Hybrid cloud**: It is a blend of private and public cloud computing, coined together by the use of technology. Where some services are available to all and some are limited to a group of consumers.

## 13.2 Internet of Things

The swift evolution of Cloud computing has driven many new technologies like Fog computing, IoT and Edge computing, and so forth. Figure 13.3 [1, 2].

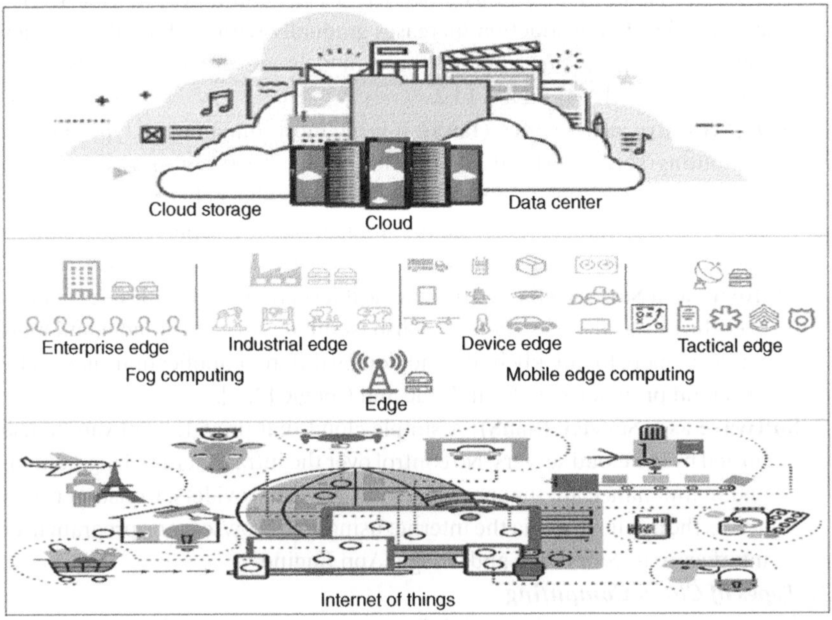


**Figure 13.3** Arising technologies in cloud computing [1–3].

These promising technologies can interoperable with the cloud or are cloud-based. However, they are summing to the complexity of this incorporation is turning into a central participant in remaking the universe of big business registering. IoT offers a complete computing solution with its advanced capabilities and many new features.

The IoT's principal objective is to engage anything so it might send and get information from and to anybody using network access whenever and from anyplace. It uses common and new technologies like networking, robotics, and sensing to allow tighter integration of items within a system. The IoT's principal objective is to engage anything so it might send and get information from and to anybody using network access whenever and from anyplace. It uses common and new technologies like networking, robotics, and sensing to allow tighter integration of items within a system.

Sensing elements, microcontrollers, communication elements, traversal features, data storage, and a network protocol are all included in IoT devices. As illustrated in Figure 13.2, the main building elements of IoT are [6–10].

- Devices/Sensors/actuators
- Gateways/Network Infrastructure
- Cloud Infrastructure
- Analytics/Processors
- User Interface

The effective functioning of the Internet of Things is ensured by these basic blocks. Smart web-enabled gadgets are incorporated with inbuilt sensors in IoT systems, which collect data from their surroundings. With the help of technologies like actuators, Radio Frequency Identification (RFID), wireless sensors [11], and satellite positioning, they are given a unique ID. Devices may occasionally connect to obtain related data. IoT Gateways then transfer this data to a cloud infrastructure. The smooth data stream and security of data transit are managed by gateways. Using AI or ML-based models, cloud applications transform such data into intelligent information. The extracted data is analyzed, and the results are used to determine what actions should be made [6, 7]. The action could be automated, with the sensors or equipment being handled without the need for human intervention.. For instance, let, the temperature sensor detects a temperature rise, the AC is immediately turned on after analysis. User interface gadgets or apps get instructions to proceed in the form of alerts, notifications, or other types of notification to the user, or if user inputs are required to complete the action. As a result, make intelligent decisions to control physical entities.

With the rise in internet-empowered smart devices, a slew of IoT applications is cropping up in practically every field. The Internet of Things has improved people's lives. Many developments, such as Google Car, a self-driving car with real-time information interchange of road, traffic, and weather conditions, have been made possible by applications with autonomous characteristics that interact with linked devices and share information. Every aspect of our lives is getting changed by the growth of IoT [6]. The fast expansion in the number of associated gadgets to the web and the huge advances in data and heterogeneous communication technologies have prompted the incredible rise of the IoT. Because of these advances, IoT systems are strongly utilized in numerous applications as exhibited underneath. Figure 13.3 outlines some of the fundamental areas and applications of IoT [7–9, 12]. The following are a couple of uses of IoT.

A) ***Area of IoT Applications***

With an increment in web empowered savvy gadgets, there are plenty of IoT applications arising each day in practically all regions. IoT has raised the quality of life. Applications with independent highlights interface with related gadgets and offer data and in this way give way to numerous developments like Google Car, a self-driving with constant data trade of street, traffic, and climate conditions [13]. All aspects of our lives are getting changed by the improvement of IoT [6]. The quick extension in the number of related contraptions to the web and the immense advances in information and heterogeneous correspondence advances have incited the staggering ascent of the IoT. In light of these advances, IoT systems are seriously utilized in numerous applications as exhibited beneath. Figure 13.3 diagrams a portion of the principal regions and uses of IoT [7–9, 12]. Coming up next are a couple of utilizations of IoT.

- **Smart Tracking of Traffic**: Traffic today is a key piece of everybody's everyday lifecycle and its issues sway society; hence, they should be tended to. IoT innovation contributes possibly in getting to the earlier data on deals jams, street conditions, an ideal course, and so on, and help with settling on ideal choices. A smart traffic observing framework benefits highlights like mishap revealing, cost assortment, burglary location, and so forth A climate versatile traffic lighting framework saves energy and ends up being practical. Brilliant city fuses a shrewd parking spot framework monitoring their accessibility and status [14].
- **Smart Homes**: Home apparatuses like A.C, fridges, TVs and clothes washers, and so on can be moved up to be savvy and permit productive energy the board. A home observing answer for the older, a power utilization checking framework for a financially savvy arrangement, and a home gatekeeper framework for assurance against burglary and infringement can be created by incorporating with IoT [15, 16].

- **Savvy Environment**: Weather estimates and regular catastrophes like floods, seismic tremors, rainstorms, cloud exploding, and so on can be ideal anticipated by IoT products. Preventive estimates like the clearing of individuals from the impacted region, or setting off salvage tasks there can be taken well on schedule. Water quality observing can likewise be cultivated by IoT, its sensors that action basic water boundaries to guarantee quality water supply [17, 18].
- **Assisted Healthcare**: Smart wearable implements can be related to the patients that can detect and report the clinical boundaries identified with patients like temperature, beat, pulse, heartbeat, and so on to the attendants and specialists. Drone ambulances are an IoT model currently in business sectors that can work with first aid kits before the specialists could arrive at the scene [19].
- **Smart Agriculture**: IoT application screens the nourishment levels of soil, light, dampness, and so forth, and can change the temperature to make the ideal nursery impact, exact watering, and preparation, hence further developing creation [17].
- **Supply Chains Coordination**: Amid the utilization of IoT applications and Radio Frequency ID innovation items in the Retail Marketplace can be effectively followed. Retailers get helped with the administration of the stock levels and computerized reordering when items leave stock and can plug shoplifting. Likewise, IoT applications possibly can engage organizations with highlights like a business gauge, reports, diagrams, and charts which can assist with making viable promoting techniques [18].

Internet is overloaded with options to choose from. A feasible, tailored and speedy solution can be rendered with the assistance of a Recommender System based on 5G technology. A basic introduction to 5G technology and an overview of the Recommender System and the significant role that the Recommender system via 5G can play in supporting and establishing the growth and advancement of the Internet of things with easily approachable solutions to everyone is discussed following sections.

## 13.3 5G Technology

Mobile access technology has witnessd a long journey from basic voice calls in first generation followed by text messages in second generation, enabled mobile internet access in third generation, till fourth generation having better internt data speed and support for apps. With order to cater new pace of technology and future needs, some of the main requirements to be met are increased capacity, better data transfer rate, lower latency and better quality of service. To meet these requirements, the cellular network architecture must be drastically improved. This laid the foundation of 5G. It is the next step in mobile network evolution as it brings faster speeds, lower latency and that supports an even more connected and technologically advanced world [19, 20].

*5G Technology:*

- Utilizes the amalgamation of high frequency radio waves.
- Incorporates newfangled technologies such as MIMO (multiple input, multiple output) and beamforming to improve signal strength and efficiency.
- Can offer significantly higher data transfer speeds than 4G.
- Condenses the time taken by the devices to communicate with each other, making interactions almost instantaneous.
- Powered enhanced Internet of Things (IoT) device compatibility and quicker video streaming.
- Provides basisi for emerging technologies such as virtual reality (VR), augmented reality (AR), and driverless cars etc.

By uniting with the power of 5G technology, a recommender system can provide just-in-time and higly interactive recommendations. Let's understand howbouts of a Recommender system.

## 13.4 Recommender System

A recommender framework is a mechanized framework to channel a few substances, that can be items, advertisements, motion pictures, recordings, melodies, or people groups from clients' e-cooperation and looks to predict the "inclination" a client would stretch out to a thing by utilizing a few calculations or computerized strategies [21]. Business applications specifically use recommender frameworks. It doesn't just work for which thing to suggest yet in addition to what the future held being positioned.

With the plenty of things/items accessible on the web, it's a seriously unwieldy errand for the client to pick or cross the right item one is searching for. Recommender frameworks not just simplify this assignment for the client yet additionally additional benefits for the organizations online to shoot up their benefits. They can offer a huge showcase of things to their client, without racking them truly.

Recommender System can

- Makes data separating simple from data over-burden over the web.
- Improves client encounters.

- Decreases search and route time to track down the ideal item.
- Builds efficiency and believability.
- Find Products, that probably won't be found in any case.
- Common gainful suggestion

B) ***Recommender System and Its Types***

A recommender framework is an electronic program that offers the dynamic client a rundown of things as proposals grounded on clients' communication with the framework. Given the calculation utilized, a Recommender framework can be ordered into Collaborative, Content-based, Demographic [21], information-based, and Hybrid frameworks [22] as in Figure 13.4.

**Collaborative Filtering (CF)**: It makes proposals because of clients' records of communications with the thing as well as equivalent choices made by different clients for comparative things independent of the thing's highlights [22, 23]. The idea driving it is, assuming two shoppers have likewise evaluated n things, there will be a high chance that they will rate different things likely.

**Content-Based**: It Makes ideas by mining the substance data of the items and clients [21, 22].

**Demographic**: Clients' segment qualities like orientation, age nation, vocation, instruction, and so forth are utilized to investigate suggestions [21, 22].

**Knowledge-Based**: It Maps the clients' express prerequisites with the arrangements accessible in the framework [21–24]. Ideas are not the aftereffect of clients' appraising history. This type is appropriate for items that are not bought habitually like houses, vehicles, and so forth.

**Hybrid**: It mixes the procedures of more than one suggestion strategy [25–27] consequently Leveraging the up-sides of a technique and beating its limits.

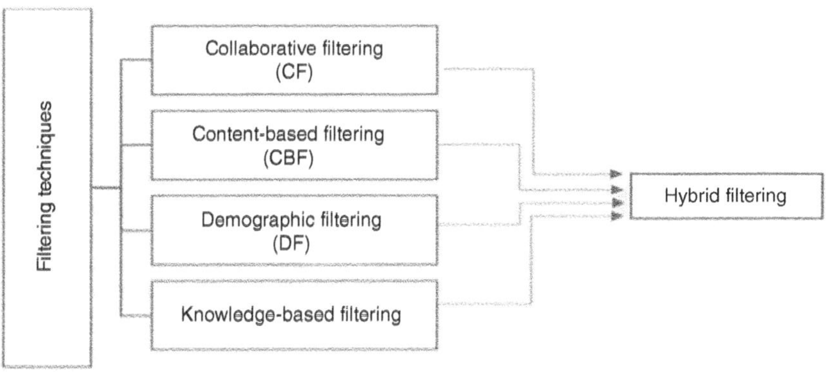


**Figure 13.4** Types of recommender system techniques [23–27].

Recommender systems, Cloud and IoT integrated with 5G have the potential to revolutionize how we interact with information, products and our overall experience online. Here's an analysis of the synergy between them:

| | **Potential benefits** |
|---|---|
| Enhanced user experience | The rapid speeds and low latency of 5G can greatly enhance recommender system performance. Instant suggestions are given to users, which is very helpful for real-time activities like gaming, streaming, and other apps. |
| Real-time context-aware recommendations | The speed of 5G makes it possible to analyse user behaviour in real-time, which produces suggestions that are more dynamic and pertinent. Imagine entering a store and getting instantaneous product recommendations based on your browsing history, location, and requirements at the moment. |
| IoT and data collection | Recommender algorithms are able to use larger amounts of data from IoT devices because to 5G's enhanced connection. Based on a wider range of user preferences and behaviours, this may result in suggestions that are more precise and tailored to the individual. |
| More comprehensive data sources | A greater range of data may be handled by 5G, including sensor data from linked devices, wearables, and smart homes. With this information, highly customised recommendations based on your particular context and surroundings may be made. Consider being able to choose a restaurant based on factors like your present location, mood, and dietary requirements. |
| Edge computing | By combining edge computing with 5G, latency may be further decreased. Recommender systems may analyse data closer to the source by utilising edge computing, which improves the recommendation's real-time quality. |
| Collaborative applications | Because 5G networks are collaborative by design, they may give rise to real-time collaborative recommender systems in which users or devices work together to improve suggestions. Users present at a live event, for instance, might affect suggestions jointly. |
| Predictive and proactive recommendations: | Recommender systems have the ability to anticipate user wants and provide pertinent suggestions for products before the user ever expresses them by analysing real-time IoT signals and historical data. Your smart home system may, for instance, proactively change the temperature based on historical preferences and current meteorological information. |
| Improved content distribution | 5G speeds up the distribution of content, making it possible to propose richer formats, such as immersive experiences, high-definition films, and real-time recommendations in AR/VR settings. |

| Potential benefits | |
|---|---|
| Enhanced user experience | The rapid speeds and low latency of 5G can greatly enhance recommender system performance. Instant suggestions are given to users, which is very helpful for real-time activities like gaming, streaming, and other apps. |
| Better offline suggestions | By using locally stored data and processing capacity, 5G's edge computing capabilities make suggestions possible even when offline. |
| Scalability and efficiency | Cloud computing offers the capacity to manage the vast amounts of data produced by Internet of Things devices and effectively power sophisticated recommender systems. It is made possible by the scalable processing and storage capabilities offered by cloud computing. Cloud-based processing and storage enable recommender systems to analyse user behaviour and train machine learning models. And 5G technology adds up lightening speed to solutions offered. |
| **Chanllenges and concerns** | |
| Security and privacy | 5G and recommender systems both need to take privacy issues into consideration. Ensuring safe and privacy-preserving recommender algorithms becomes increasingly important as more data is delivered and processed. Privacy problems arise when large volumes of personal data are collected and used. Strong data anonymization and user permission procedures must be put in place. |
| Ethical contemplations | Discrimination may be sustained by algorithmic bias in suggestions. Ensuring diversity and justice requires careful planning and oversight. |
| Infrastructure cost | The development and upkeep of 5G infrastructure demand large financial outlays. |
| Digital divides | A digital gap may result from unequal access to 5G networks, cloud computing, and smart gadgets. It is necessary to address equitable access. |

Figure 13.5 show the prevailing and the estimated trends of the above technologies summarized according to the McKinsey Global Institute report, IDC report and GSMA Intelligence [28–30].

Recommender system, cloud, IoT, and 5G integration can result in a comprehensive and potent ecosystem. It allows for seamless communication across a variety of apps, effective data processing, and personalised, real-time suggestions. To guarantee the effective and moral deployment of this integrated system, concerns like security and privacy must be properly handled.

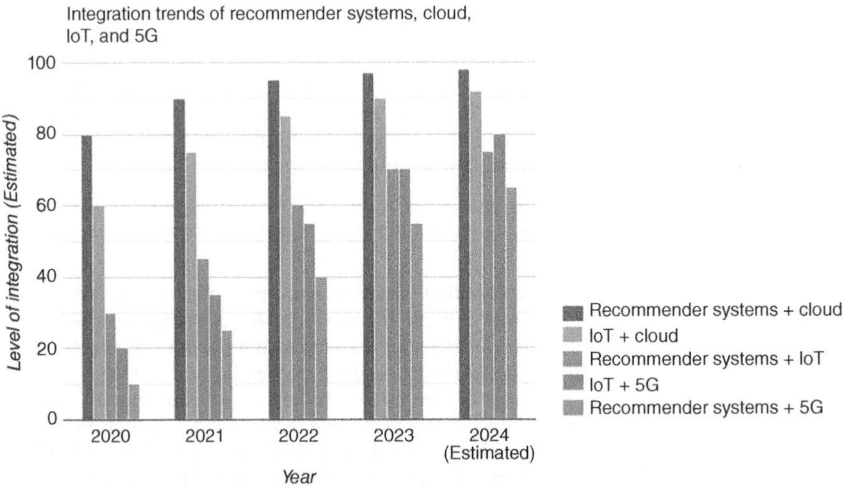


**Figure 13.5** Trends of integration of recommender system, cloud computing, IoT and 5G.

## 13.5 Conclusion

Owe to forward-thinking technologies like Cloud Computing, the Internet of Things (IoT), and the revolutionary potential of 5G connection, the modern world has transformed into a more intelligent state. Cloud computing fosters productivity at a cost-effective rate by offering IT-based services in a variety of formats. IoT has made technology more approachable by introducing intelligence at the granular level. The abundance of alternatives and selections has, however, made it difficult for this technology to be adopted smoothly.

The addition of 5G technology improves things even further by providing previously unheard-of speeds, minimal latency, and simultaneous connections to several devices. In addition to speeding up data transfer, 5G's high speed and low latency guarantee real-time communication and responsiveness in Internet of Things applications. The interplay of 5G, IoT, and cloud computing produces a potent environment in which data is reliably and quickly transferred in addition to being easily available.

## References

**1** Mell, P. and Grance, T. (2011). The NIST-National Institute of Standards and Technology- definition of cloud computing. *NIST Special Publication 800-145*, p. 7.

2 Sosinsky, B. (2011). *The Book You Need to Succeed! Cloud Computing*, 1ste. Indianapolis/Indianapolis, Indiana: Indiana Published/Wiley Publishing, Inc.

3 Borko Furht, A.E. (2010). *Handbook of Cloud Computing*. Boston, MA: Springer US.

4 Microsoft Corporate Author (2016). What is cloud computing? A beginner's guide| Microsoft Azure. Cloud Comput., p. 1, [Online]. Available: https://azure.microsoft.com/en-in/overview/what-is-cloud-computing/#cloud-computing-models (accessed 22 July 2020).

5 Velte, T., Velte, A., and Elsenpeter, R. (2009). *Cloud Computing, a Practical Approach*. USA: McGraw-Hill, Inc.

6 Patel, K.K., Patel, S.M., and Scholar, P.G. (2016). Internet of Things-IoT: definition, characteristics, architecture, enabling technologies, application & future challenges. *International Journal of Engineering Science and Computing* 6 (5): 1–10. https://doi.org/10.4010/2016.1482.

7 Kumar, N.M. and Mallick, P.K. (2018). The Internet of Things: insights into the building blocks, component interactions, and architecture layers. *Procedia Computer Science* 132: 109–117. https://doi.org/10.1016/j.procs.2018.05.170.

8 Internet of Things (IoT) - Part 2 (Building Blocks & Architecture). https://www.c-sharpcorner.com/UploadFile/f88748/internet-of-things-part-2/ (accessed 18 March 2022).

9 IoT Explained – How Does an IoT System Actually Work? | by Calum McClelland | IoT For All | Medium. https://medium.com/iotforall/iot-explained-how-does-an-iot-system-actually-work-e90e2c435fe7 (accessed 18 March 2022).

10 Serbanati, A., Medaglia, C.M., and Ceipidor, U.B. (2011). Building blocks of the nternet of Things: state of the art and beyond. In: *Deploying RFID - Challenges, Solutions, and Open Issues*. https://doi.org/10.5772/19997.

11 Atzori, L., Iera, A., and Morabito, G. (2010). The Internet of Things: a survey. *Computer Networks* 54 (15): 2787–2805. https://doi.org/10.1016/j.comnet.2010.05.010.

12 Gubbi, J., Buyya, R., Marusic, S., and Palaniswami, M. (2013). Internet of Things (IoT): a vision, architectural elements, and future directions. *Future Generation Computer Systems* 29 (7): 1645–1660. https://doi.org/10.1016/j.future.2013.01.010.

13 Macedo, D., Guedes, L.A., and Silva, I. (2014). A dependability evaluation for Internet of Things incorporating redundancy aspects. *Proc. 11th IEEE Int. Conf. Networking, Sens. Control. ICNSC 2014*, pp. 417–422. https://doi.org/10.1109/ICNSC.2014.6819662.

14 K. Tupe, V. Jandrao, S. Chappe, P. Bhusalwad, and P. E. S. College, "Child health care monitoring using sensor technology," *International Research*

*Journal of Engineering and Technology*, vol. 4, no. 3, pp. 3–6, 2017, [Online]. Available: https://irjet.net/archives/V4/i3/IRJET-V4I389.pdf.

**15** Siraj, S. (2019). ARCTIC: An IoT-Based System for Child Tracking in Day Care.

**16** Nandini Priyanka, M., Murugan, S., Srinivas, K.N.H. et al. (2019). Smart IOT device for child safety and tracking. *International Journal of Innovative Technology and Exploring Engineering* 8 (8): 1791–1795.

**17** Liu, Y., Yang, Y., Lv, X., and Wang, L. (2013). A self-learning sensor fault detection framework for industry monitoring IoT. *Mathematical Problems in Engineering* 2013: https://doi.org/10.1155/2013/712028.

**18** Bhatnagar, A., Shukla, S., and Majumdar, N. (2019). Machine learning techniques to reduce error in the Internet of Things. *Proc. 9th Int. Conf. Cloud Comput. Data Sci. Eng. Conflu*. pp. 403–408. https://doi.org/10.1109/CONFLUENCE.2019.8776619.

**19** Gupta, A. and Jha, R.K. (2015). A survey of 5G network: architecture and emerging technologies. *IEEE Access* 3: 1206–1232. https://doi.org/10.1109/ACCESS.2015.2461602.

**20** Shafi, M. et al. (2017). 5G: a tutorial overview of standards, trials, challenges, deployment, and practice. *IEEE Journal on Selected Areas in Communications* 35 (6): 1201–1221. https://doi.org/10.1109/JSAC.2017.2692307.

**21** Sarı, A., Tosun, A., and Alptekin, G.I. (2019). A systematic literature review on software fault prediction and fault tolerance in software engineering. *Journal of Systems and Software* 153: 200–219. https://doi.org/10.1016/j.jss.2019.04.027.

**22** Pooja Paramshetti, D.A.P. (2014). On software defect prediction using machine learning. *Journal of Applied Mathematics* 2014 (12): 1–8.

**23** Francesco Ricci, Lior Rokach, Bracha Shapira, Paul B. Kantor. *Recommender Systems Handbook*, 2011, https://doi.org/10.1007/978-0-387-85820-3.

**24** Sharma, S., Gupta, K., and Gupta, D. (2021). Recommender system: a bibliometric analysis. *IOP Conference Series: Materials Science and Engineering* 1022 (1): https://doi.org/10.1088/1757-899X/1022/1/012057.

**25** Alhijawi, B. and Kilani, Y. (2020). A collaborative filtering recommender system using genetic algorithm. *Information Processing and Management* 57 (6): 102310. https://doi.org/10.1016/j.ipm.2020.102310.

**26** Sielis, G.A., Tzanavari, A., and Papadopoulos, G.A. (2014). Recommender systems review of types, techniques, and applications. In: *Encyclopedia of Information Science and Technology*, 7260–7270. https://doi.org/10.4018/978-1-4666-5888-2.ch714.

**27** Burke, R. (2002). Hybrid recommender systems: Survey and experiments. *User Modeling and User-Adapted Interaction* 12 (4): 331–370. https://doi.org/10.1023/A:1021240730564.

**28** https://www.gartner.com/en/information-technology/insights/top-technology-trends

**29** https://blogs.idc.com/2022/12/12/idc-futurescape-worldwide-future-of-industry-ecosystems-2023-predictions-draft/

**30** https://data.gsmaintelligence.com/research/research/research-2023/global-mobile-trends-2023

# 14

# Confluence of Cellular IoT and Data Science for Smart Application using 5G

*Shruti*[1,2] *and Shalli Rani*[2]

[1] *PG Depatment of Information Technology, Goswami Ganesh Dutta Sanatan Dharma College, Chandigarh, India*
[2] *Chitkara University, Institute of Engineering and Technology, Chitkara University, Rajpura, Punjab, India*

## 14.1 Introduction

According to study, world's total population will be around 9.1 billion by 2050 and the global demand for agricultural production will also increase by 70%. To meet up with the demands of increasing population and to increase the agricultural production, the government along with agricultural industry is turning towards Internet of Things (IoT) and data analysis methods. IoT is the future that comes with a rule, "Anything can be connected to anything". It is an interconnection of humans, animals, objects, machines, and computer devices. Every connected device in IoT network is given unique ID that has an ability to transfer data without any kind of interaction with humans or computers. Now a days, a variant of IoT known as Cellular IoT is gaining popularity. It uses cellular networks to provide connectivity to the IoT devise. In traditional cellular communication, the focus is on connecting smartphones and other computing devices but with the advent of IoT, there is a growing need for a reliable and scalable network infrastructure to connect multiple devices beyond just phones and computers. Therefore, Cellular IoT encourages existing cellular technologies to provide connectivity for a wide range of IoT devices. There are various technologies within cellular IoT like LTE-M (Long-Term Evolution for Machines), NB-IoT (Narrowband IoT), 5G, etc. that specifically caters the needs of IoT devices. These technologies offer improved power efficiency, scalability and bandwidth compared to traditional cellular technologies. Some of the key features of cellular IoT includes:

1) **Wide Coverage Area:** Cellular networks have wide coverage, making them suitable for connecting IoT devices in various locations, including remote or rural areas.

*Current and Future Cellular Systems: Technologies, Applications, and Challenges*, First Edition.
Edited by Garima Chopra, Suhaib Ahmed, and Shalli Rani.

2) **Scalability:** Cellular networks can handle a large number of devices simultaneously, making them scalable for IoT applications with various connected devices.
3) **Reliability:** Cellular networks are designed to provide reliable and stable connectivity, ensuring consistent communication between IoT devices and the server.
4) **Security:** Cellular networks integrate security features like encryption, to protect data while transmitting it between IoT devices and the network.
5) **Standardized Protocols:** Cellular IoT uses standardized communication protocols to make it easier for device manufacturers to ensure compatibility and interoperability.

While on the other hand, data science is also becoming a vital part of our industries. As a large amount of data is produced, the companies have started moving towards the implementation of data science techniques to expand the business and increase customer satisfaction. It can be defined as a study of large amount of data that with the help of tools and techniques is able to find hidden patterns, derive useful information and make decisions in businesses. Data can be from any source and in any format. Machine learning in data science can be used for building predictive models. One of the initial steps in data processing is data pre-processing. Here, raw data collected from various sources is converted into meaningful one. This step is done by data scientists so that they get appropriate data for what they want to do. Data pre-processing goes through some phases like data cleaning, data amalgamation, data conversion, data reduction and data discretization. IoT gathers data that are one of a kind and capable of associating with the Internet. Data can be anything that can interact with internet and with each other so that it can communicate. Communication is irrespective of the capability of device about what it can do or not. Some of the reasons why devices communicate with each other in IoT environment is as follow:

- For communication
- For catching of data
- Storing data
- For processing

Many difficulties are faced while acknowledging IoT. Some of them are design requirements of the device, privacy of the data and device stimulation so that they can communicate with each other. As the use of IoT based applications is increasing, data volume, speed of data communication and reliability also need to be improved. This huge blast in data need a better foundation along with a greater number of intelligent data scientists to handle it. Ultimately, IoT is not only about sourcing data but to extract knowledge from this data. And data science can contribute to make this thing happen and produce more intelligent systems. Using different techniques to find hidden patterns, data science consists of wide range of algorithms that can be applied in various domains. Some of the methods of smart farming that includes IoT sensors and data processing procedures for future predictions and data storage are already being used by farmers. The paper reviews different IoT platforms and sensors that can be used by farmers. Different methods of data processing and how data science can be helpful in agriculture are also analysed. Thus, the main focus is how cellular IoT by using 5G technology, and data science can benefit agricultural field in best possible manner. A model, which is a confluence of cellular IoT, data science and blockchain is also discussed which will provide security to the farmers.

On the other hand, blockchain is a decentralized and distributed digital technology that securely records and verifies the transactions across a network of computers. A chain of blocks containing a list of transactions is there, once a block is filled with transactions, it is linked to the previous block, creating a chain of blocks. The information contained in a blockchain is secure, transparent, and resistant to modification. It eliminates the need for a central authority. The integration of blockchain with cellular IoT and data science unlocks a scope of secure and transparent possibilities for smart applications. Blockchain's decentralization ensures the integrity of data generated by cellular IoT devices, offering tamper-proof records which are crucial for applications that requires data accuracy. Furthermore, it also enhances security in edge computing environments and provides transparency in data science models, ensuring trustworthy decision-making. This convergence ensures innovative and secure smart applications across industries, from agriculture to smart cities.

The paper is organized in various sections. Section 14.1 gives introduction on IoT, cellular IoT, data science and their prospects in agriculture. Section 14.2 addresses detail insight into cellular IoT and data science, how they both can be related and differentiated from each other. Research problems in data science for cellular IoT is discussed in Section 14.3. Section 14.4 is all about different types of sensors that are available in smart farming. It also discusses what type of data is collected by these sensors. Related work in the field of agriculture using IoT, cellular IoT and data science is discussed in Section 14.5. Section 14.6 explains about data science in agriculture. Challenges faced by cellular IoT applications in data science are mention in Section 14.7. A blockchain model including cellular IoT and data science that can be used in farming and how it can be used to provide data security to farmers is mentioned in Section 14.8 and Section 14.9 concludes the paper.

## 14.2 Data Science and Cellular IoT

Over the years, IoT has become one of the important technologies. It contains a network of sensors, machines, end-users, and data centres. IoT popularity is rising day by day due to increase in its productivity, better services provided, creating new products, and letting us know more about the world around us. Data science is a key for the growth of IoT. Therefore, we can say that data acts as a fuel for IoT and data science turns the fuel into asset. Data science can use its variant cellular IoT massively for many real time applications associated with data collection and computations as per following [1]:

1) **Cognitive Computing:** manages hard to reach districts by business understanding and assessments.
2) **Processing in Real-Time:** When data is processed in real time, a continuous data generation takes place. Real time data processing helps to make quick analysis and permits to act within seconds in case of need.
3) **Time Series Analysis of Data:** In this data is arranged according to time.
4) **Geo-spatial Data Analysis:** Here, data is analysed according to the geographical area. But this type of analysis face challenges like advancement in finances and rising innovations.
5) **Machine Learning:** Machine learning models are unconventional when dealing with organized data. In case of data like pictures, voice, videos and recordings, a lot of time is required, and they are generally weak and not adaptable.
6) **Edge Computing:** It encourages to cut down costs that otherwise is very high in case of data set exchange over the system.

Traditional data science deals with data accumulation and its association without worry about data expiry and timespan. Whereas, cellular IoT have to examine and understand data depending on the response time of the application especially in case of real-time applications. Traditional data and cellular data have difference between them that can be discussed in Table 14.1.

Categorizing data science solutions for cellular IoT systems, the first thing is the source of data. Data can be produced from various areas like sensors, public/private sectors, intelligent systems, etc. After sourcing, system components are there that look after processing of data, maintenance of data, data conveyance and data analysis. Third classification is of data science technologies which can be related to IoT, artificial intelligence, machine learning and deep learning. Then comes the elements that efficiently supports IoT: Input data, data maintenance rules, computational devices required for processing and output data. Lastly, the data analysis which can be classified into expressive, prognostic, and rigid.

**Table 14.1** Difference between traditional data and cellular IoT data.

| Features | Traditional data | Cellular IoT data |
|---|---|---|
| Type of connection | Wired or wireless | Cellular network |
| Device range | Limited to a network | Extended due to cellular coverage |
| Scalability | Difficulty in handling large data simultaneously | Scalable to accommodate large number of IoT devices |
| Data rate | Vary, depends on the type of network | 5G offers high data rate |
| Latency | High | 5G offers low latency |
| Infrastructure | Relies on wired network or Wi-Fi | Cellular network provides more coverage |
| Domain | Traditional computing, communication and local networking | IoT application like smart cities, healthcare, agriculture, etc. |

A large amount of data created from different arrangements whether organized, unorganized, or semi-organized in cellular IoT is carefully taken care of and arranged by different data accumulation techniques of data science (Fig. 14.1). This data is further consumed and stored using data stacking. It is then prepared and investigated using big data examination. Cellular IoT can self-govern itself by refining and keeping its data up-to-date, then only data science and machine learning can be helpful. Combination of both has been seen as an improvement in human life. Data science systems requires the data to be properly characterized after which calculation methods can be applied to provide distinctive data quality.

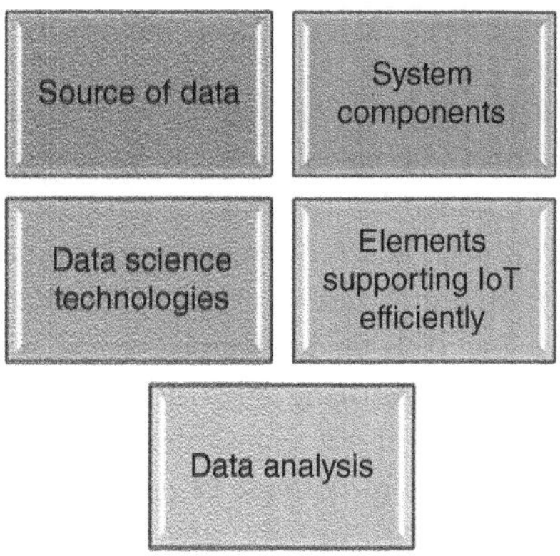


**Figure 14.1** Data science solutions to be used for IoT system.

The data science for cellular IoT and traditional data science can be different from one another in following ways:

- Ongoing processes and IoT
- Capability of IoT and artificial intelligence
- Machines on which they will be working
- Classification according to real-time
- Data accumulation in real-time
- Connections
- Data collection by sensors

Data science combines statistics, data analysis and other approaches to know and examine the authenticity of the occurrence of anything through the data. And with machine learning this area is becoming more satisfying. Programming languages like python, java, SQL, R, Scala, and Julia can be used to understand and use data science.

The integration of 5G cellular IoT with data science enhances smart applications across various domains like smart cities, healthcare, smart farming, etc. The high-speed, low-latency connectivity of 5G enables seamless data transmission from IoT devices, enabling real-time insights. Data science, with its analytical ability, processes and interprets the incoming data, extracting valuable patterns and trends. In farming, data processed by data science algorithms, provides actionable insights into soil conditions, crop health, and weather patterns. Farmers can make informed decisions, optimizing irrigation, crop management, and resource allocation. This enhances precision in agriculture, improves yield, sustainability, and overall farm efficiency.

## 14.3 Research Problems in Data Science for Cellular IoT

In data science for cellular IoT, researchers face several challenges. The primary concern being ensuring the security and privacy of data transmitted over cellular IoT networks. Developing a robust mechanism to protect sensitive information is crucial and requires encryption and authentication solutions to mitigate risks of unauthorized access or cyberattacks. Scalability emerges as another problem, as IoT devices are widely spread within the cellular networks. The challenge lies in optimizing algorithms to process diverse data streams from numerous devices without compromising efficiency. Energy efficiency is also a critical concern in the context of cellular IoT devices that often operate on limited power sources. Researchers aim to develop algorithms, focusing on minimizing the computational and communication energy consumption of IoT devices to extend their operational life.

Ensuring the quality and reliability of data is an ongoing research problem. Factors such as sensor accuracy, calibration, and potential data anomalies must be addressed to guarantee the dependability of insights derived from cellular IoT data. Managing dynamic network conditions is also important. A data science solution capable of dynamically adjusting to changes in the cellular network environment is required. This adaptability optimizes data processing and analytics based on evolving network conditions and device connectivity. Fault tolerance and resilience are essential considerations for maintaining the functionality of cellular IoT applications. Strategies to ensure the continued operation of data science applications even in the presence of network disruptions or device failures is must. For trustworthy models, researchers should examine explainability in data science for cellular IoT. It is crucial for critical applications where transparent decision-making based on IoT data is important. Various IoT devices create a stream of data and data analysts can remove unwanted data using some procedures. This data stream then needs to be understood and examined for further processing that stimulates the data science. Machine learning plays an important role along with data science to handle this vast cellular IoT data.

Addressing these research problems in data science for cellular IoT will contribute to the development of secure, efficient, and robust solutions for the evolving landscape of IoT applications. And in this paper, we will solve most of these problems by developing an efficient blockchain model that includes cellular IoT and data science, that can be used in farming providing data security to the farmers.

## 14.4 Sensors in Cellular IoT Smart Farming

After growing crops, its maintenance require adequate amount of environmental conditions. In smart farming, sensors play an important role to look after all the parameters required for the growth and health of the crops [2]. Sensors can be of various types depending on methods, material type, sensing capability and their application. Some of the sensors that can be used in agriculture are discussed below:

1) **Temperature Sensors:** These types of sensors include soil, crop and air temperature sensors. Soil temperature sensors plays an important role to control microbiological processes like germination, photosynthesis, mineralization, root growth, and decomposition that will help in the growth of the crop [3]. $ECH_2O$, Hydra probe II, EC250 and MP406 are few sensors that can be

used to measure temperature of the soil during farming. On the other hand, maintaining crop temperature helps in productivity and quality of the crop. By sensing crop temperature, we can predict the yield of the crop. Infrared images of the leaves are used to get temperature measurements. LT-1 M senses leaf temperature and do predictions. Air sensors senses rate of precipitation, evaporation, humidity, and speed of the wind. 107-L and SHT-71 are example of air temperature sensors.

2) **Humidity Sensor:** Humidity sensors checks for moisture and air temperature because imbalance in these will affect photosynthesis. CI-340 measures $CO_2$, humidity, transpiration, and photosynthesis.
3) **Moisture Sensor:** These sensors detect the amount of water/moisture in the soil that will help to manage irrigation. MP406 is a moisture sensor that checks soil and take continuous measurements to maintain the water contents in the soil.
4) **Pesticide Sensor:** As the continuous use of pesticide damages the crop. Pesticide sensors are used to sense the amount of pesticide used on crops so that they can be protected from any kind of damage.
5) **Rain Sensor:** Heavy rainfall can be damaging to the crop. So, to protect crops from any negative impact of rain rain-sensors are used.

There are few more other types of sensors that can be helpful for precise agriculture. Gas sensors can be used to check the number of harmful gases like methane in greenhouses. Motion sensors can be used by farmers to detect the motion of animal or any unwanted object in their fields which may damage the crop. Pressure sensors checks for air or atmospheric pressure. UV sensor, infrared sensor, electromagnetic sensor, optical sensor, mechanical sensors, and others are few types of sensors that can be helpful in agriculture, Table 14.2 below discuss few of those [4].

The relation between various environmental factors studied above is discussed using linear regression model. Linear regression illustrates how an increase/decrease in one factor affects the other, it is an estimation of relationship between 2 or more variables. Linear regression model here is applied on the data obtained from the dataset to find the relationship between the factors like humidity, pH level, air temperature, etc. A simple linear regression equation is written as:

$$\alpha = \beta_0 + \beta_1 x_1 \tag{14.1}$$

where $\alpha$ = predicted value, $\beta_0$ = regression constant, $\beta_1$=quantifies the effect of independent variable on dependent variable and $x_1$ = independent variable sample data. $\beta_0$ is obtained from data representing y-intercept and $\beta_1$from slop of the line.

The graph below (Figure 14.2) shows the relation between air temperature and humidity.

**Table 14.2** Types of sensors used in agriculture.

| Sensor | Example | Parameters it measures |
|---|---|---|
| Temperature | Soil temperature<br>• Pogo portable soil sensor<br>• EC250<br>• THERMO200 | • Soil temperature and soil moisture<br>• Soil temperature, soil moisture, salinity and conductivity<br>• Soil and air temperature |
| | Crop temperature<br>• PTM-48A<br>• 237 leaf wetness<br>• LT-1 M | • Crop temperature, moisture, wetness and $CO_2$<br>• Crop temperature, moisture and wetness<br>• Crop temperature |
| | Air temperature<br>• CM-100<br>• HMP45C<br>• XFAM-115KPASR | • Air temperature, humidity, pressure, and wind speed<br>• Air temperature, humidity, and pressure<br>• Air temperature, humidity, and pressure |
| Humidity | • DHT11<br>• AM2321 | • Air humidity and temperature<br>• Air humidity and temperature |
| Moisture | • EC250<br>• MP406 | • Soil temperature, soil moisture, salinity, and conductivity<br>• Soil moisture and temperature |
| Rain | Raindrop sensor | Rain |

## 14.5 Related Work

Researches done on IoT platforms supporting farmers in agricultural development like selecting crop according to terrain, predictions, increase in crop production, monitoring and weather forecast are discussed in this section. A lot of work has been done in IoT's smart farming so to get insight of those in agricultural field, a survey is needed where solutions and sensors that has been used in agriculture in past few years are studied. A platform called Farmbeats was developed in [5]. It collects data from cameras, drones, sensors; and uses IoT based weather-aware design to inform issues related to connectivity in the farm. It is also capable of sending high bandwidth videos and images captured by drone to the cloud. To save battery life of the drones a path planning algorithm is also incorporated. In 2017, an IoT model for agricultural areas like ecological monitoring and aquaculture

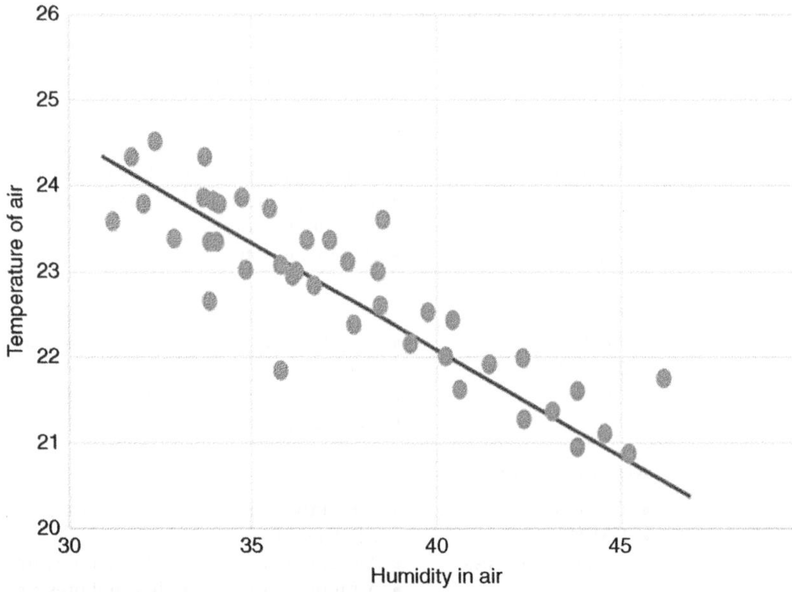


**Figure 14.2** Relation between air humidity and air temperature.

was developed [6]. Researchers can use it to develop prototypes for high-level applications like smart soil fertilization, disease forecasting, and detection. Sehan Kim et al. invented FaaS (Farm as a Service) using IoT that provides high-level application services and technology stability for agricultural environment [7]. A strawberry infection prediction system is used to evaluate the performance. N. Ahmed et al. proposed a scalable network architecture was that utilized Wi-Fi based long distance networks and fog computing for connecting rural areas in [8]. The IoT-based control system was intended specifically for farming applications in rural areas. Wen-Liang Chen et al. developed AgriTalk [9]. Farmers can access the resources of AgriTalk using any smartphone without the need of installing any application. In 2020 again, Wen-Liang Chen et al. developed an IoT device where sensors provide non-image data that Convolutional Neural Networks (CNN) use to recognize patterns [10]. Reference [11] proposes a blockchain-based fish farm platform for fish farmers for secure storage of farm data. Walmart and IBM used the Hyperledger-fabric framework to develop a blockchain-based food traceability system that boosted trust across the food supply chain [12]. To integrate big data analysis and IoT, Zhihan Lv et al. uses wireless relay cooperative transmission technology in [13]. The system performance is improved by using a multi-hop path and other approaches to lower the outage probability and bit error rate of the system. It serves as a testing ground for the development of IoT technologies. Recently, cities are undergoing massive technological and operational changes, and data science is helping in driving this transition. A scheme was proposed in which city data collected from various sources is used for insights and find hidden correlation to improve decision-making processes and provide better and more intelligent services to citizens [14].

5G is an emerging trend in cellular networks that has the potential to create new interfaces for the networking components. It plays a major role in connecting a number of users and provides a smarter and faster way of communication. It basically reduces the load and stabilizes the technical solutions, improves quality of service, provide higher bandwidth and neglects any kind of interference [15]. Aggarwal et al. [16], in their work focused on the integration of various technologies. Data mining with 5G IoT helped in making simple decision and solve different challenges, while machine learning and artificial intelligence helped in data extraction. The work also explores big data analytic which is considered crucial for mobile networks. It is expected that by 2030, there will be a massive increase in the overall network traffic and dependency on wireless communication and cellular network like 5G is ready in existence to support this huge volume of traffic. Ajmani et al. [17] represented a study of 5G technology including its potential applications and its use in the agricultural sector.

Whereas Murugamani et al. [18] in there represented a method to monitor the soil quality and prevent cotton leaf from diseases. Artificial intelligence was used for identification and classification of leaf diseases, this information is further forwarded to the farmers through an Android application. The app is further capable of displaying parameters like soil moisture, humidity, and temperature. The proposed method uses SVM algorithm to deliver best accuracy in detecting various diseases and proves its efficiency in the detection and control by improving cultivation for the farmers.

## 14.6 Data Science for Agriculture

Data science analyses and processes data using techniques like artificial intelligence, machine learning and data mining [19] which can be helpful to overcome challenges faced by agricultural industry. A large amount of data is generated by the sensors deployed in fields for operations like monitoring, weather forecast, disease detection and so on. This data is then transformed and processed that can be used for the management of farms. Multiple sensors, including soil moisture sensors and weather stations, provide real-time data on environmental conditions. Data science algorithms then process this information helping farmers to take actionable insights. These decision support systems help to optimize certain farming operations like irrigation schedules, fertilization plans, and pesticide application. By adapting these activities to the specific needs of different areas within a field, farmers can enhance resource efficiency and crop yield. Crop monitoring can be revolutionized by combining remote sensing technologies and data science. Satellite and drone images can capture detailed information about crop health, growth patterns, and potential issues. Using machine learning models, these datasets are analyzed to provide farmers with predictive understandings. It empowers farmers to make informed decisions regarding crop management, allowing timely interventions and the prevention of potential issues.

Data science applications and methods for agriculture used in past few years are studied in this section. In [20], author set up an image recognition system following IBM Foundational methodology. It was used for blueberries cultivation at its rooting stage. Agro 4.0 was developed using data science in [21] for its use in farming. It helps farmers to manage and analyse factors like fertility of the soil. An IoT based intelligent equipment to be used in agriculture was designed by Fan-Hsun Tseng et al. in [22]. It consists of sensors, power storage and web platform that is used to study various environmental conditions. In all, data science in agriculture is not just a technological advancement but also represents how farming can be improved. By connecting the power of data, farmers can make evidence-based decisions, reduce environmental impact, and optimize resource utilization, ultimately contributing to a more sustainable and productive agricultural sector.

## 14.7 Challenges Faced by Cellular IoT Application in Data Science

Combination of cellular IoT and data science creates a huge amount of data. Data created from different devices and sensors in this arrangement is taken care of and any required change is made by the variety of Big Data accumulation methods. The collected data is prepared in such a way so that it can perform any function that speeds up the process with the help of data science analysis. Some of the challenges faced by cellular IoT in data science are discussed below (Table 14.3):

1) **Storing Data:** As a large amount of data is generated by cellular IoT, storing data is a big challenge. Along with this, increased processing speed increases amount of data created which required more storage space so that more data can be analysed.
2) **Complex Computations:** The speed with which the size of the data is increasing, the current instruments may not be that effective in data processing.

**Table 14.3** Challenges and actions to be taken.

| Challenge | Action to be taken |
|---|---|
| Storing data | Data is stored and analysis is done,<br>Data can be stored in hard drives,<br>send for processing further |
| Complex computations | Data Warehouses and Data centers can be used,<br>Discharging data that is of no use,<br>vulnerability |
| Data visualization | For data adaptability and security,<br>Done when some issue arises,<br>While associating datasets |
| Poor data quality | Combines industry<br>principles and ceaseless irregularity |
| Data quantity | Large amount of data may cause issues that can affect the significant advancement |
| Various data formats | Different data formats can also cause issued in data analysis and processing |
| Speed | Irrespective of the adaptability of the cloud,<br>Not required in situation where lots of information is to be handled with time |

Data confirmation, data organization, data conservation, data recovery and reporting are complex activities related to data.

3) **Data Visualization:** The most crucial test for data science is how it provides security and adaptability to its data. Data adaptability is also important as some issues may emerge during execution. On the other hand, data representation takes cares that data is presented in a proper manner. Now a days, data analysts are more focused on preparing and examining the data.
4) **Poor Data Quality:** Manual data is usually avoided as far as possible to keep away mistakes like incorrect spelling and typographical errors. A uniform data prepared by spending time may help to prevent any re-work on the data again.
5) **Data Quantity:** A large amount of data can cause issues. In cases like this, data determination procedures can be helpful to remove unwanted/extra data.
6) **Various Data Formats:** Presence of different data formats is also challenging as data collected by cellular IoT devices makes data analysis and management difficult for data science.

7) **Speed:** Most of the processing takes place on cloud or data centres, therefore regardless of adaptability of cloud, it may not be suitable in conditions where the amount of data to be handled increases gradually.

The use of sensos and various cellular IoT based applications is increasing in agricultural sector. Along, with this several data mining techniques are also been used to process a large amount of data collected from fields. But as a huge amount of data is involved, the issues related to security and privacy is also raising with it and this leads to main challenge in agricultural field. Therefore, data science along with cellular IoT will help to overcome this problem so that farming can be seen as a flourishing field.

## 14.8 Proposed Model and its Discussion

We have discussed different types of sensors that can be used in farming and the significance of data science in smart farming. Various challenges that can be faced by smart farming is also studied in the paper. Based on the study done and inspired by the security prospects of blockchain technology, we can conclude that combining cellular IoT, data science and blockchain can improve the overall production of the crop. Along with this, it will help in empowering farmers. Farmers in order to increase their economy and get the best for them, should get themselves educated with these technologies.

Here, a cellular IoT platform that is using 5G network is designed based on blockchain technology with data modelling techniques that can be used for digitizing farming which can be used by small scale farmers. Main components used in this framework are sensors, blockchain network and end-users. The data of the crop is collected using sensors like moisture sensor, soil temperature sensor and crop temperature sensor whereas sensors like air temperature, rain sensor and humidity sensor can be used to check weather conditions. For communication, devices use 5G network to transfer data fast and securely. The data collected by sensors is forwarded for data modelling and its results along with datasets are then stored inside the blockchain framework securely. Only the users who are authenticated are allowed to perform any kind of operation on this data. The blockchain network model is explained using six farms that are of small scale and each one is equipped with IoT devices. Farmer and stakeholders are the end-users in this scenario. The model supports farmers in rural area in the way that they can interact directly with agricultural experts and take their advice instead of being exploited by some middle-men. The framework in diagrammatic form is shown in Figure 14.3 below.

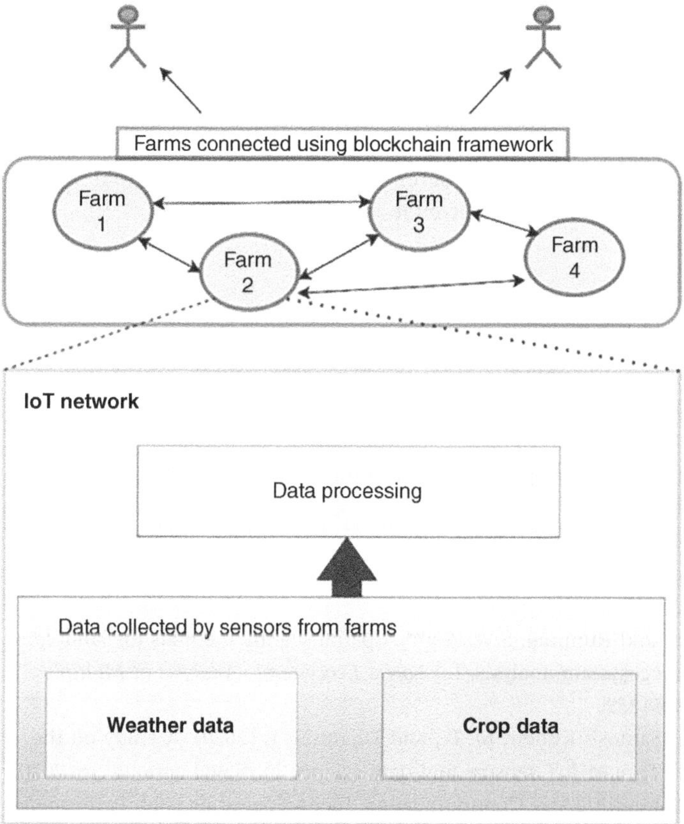


**Figure 14.3** Proposed framework to be used in smart farming.

## 14.9 Conclusion

Ultimately, cellular IoT is not only about sourcing data but to extract knowledge from this data. And data science can contribute to make this thing happen and make system more intelligent. Using different techniques to find hidden patterns, data science consists of wide range of algorithms that can be applied in various domains. Increasing population and climate change is forcing us to find techniques and methods to address agricultural issues.

Digitization in agriculture has helped farmers by providing sustainable solutions irrespective of their geographical area. The paper discusses various sensors and environmental conditions that can be taken as input for agriculture related applications, cellular IoT and data processing. A blockchain based cellular IoT framework has been proposed in the paper that will help to address security issues that may occur in smart farming due to large amount of data. This will also help the future researchers working on cellular IoT and in agricultural industry to gain from it. The future prospect holds improving the security of the smart farming using the latest blockchain technology so that efficiency can be increased much more that the one discussed here. Improvement in throughput and less computational overhead is also expected in future.

## References


1 Devi, M., Dhaya, R., Kanthavel, R. et al. (2020). Data science for internet of things (IoT). Smys, S., Senjyu, T., Lafata, P. (eds) Second International Conference on Computer Networks and Communication Technologies. ICCNCT 2019, Lecture Notes on Data Engineering and Communications Technologies, vol 44. Springer, Cham.

2 Singha, N. and Singh, A.N. (2020). Odysseys of agriculture sensors: Current challenges and forthcoming prospects. *Computers and Electronics in Agriculture*, Elsevier 171.

3 Waring, R.H. and Running, S.W. (2007). Spatial scaling methods for landscape and regional ecosystem analysis. In: *Forest Ecosystems: Analysis at Multiple Scales*, 3e, 225–259.

4 Johnson, N., Santosh Kumar, M. B., and Dhannia, T. (2020). A study on the significance of smart IoT sensors and data science in digital agriculture. 2020 Advanced Computing and Communication Technologies for High Performance Applications (ACCTHPA), Cochin, India (02-04 July 2020), pp. 80–88. IEEE.

5 Vasisht, D.,Kapetanovic, Z., Won, J. et al. (2017). FarmBeats: an IoT platform for data-driven agriculture. 14th USENIX Symposium on Networked Systems Design and Implementation, Boston USA (27–29 March 2017).

6 Popović, T., Latinović, N., Pešić, A. et al. (2017). Architecting an IoT-enabled platform for precision agriculture and ecological monitoring: a case study. *Computers and Electronics in Agriculture*, Elsevier 140: 255–265.

7 Kim, S., Lee, M., and Shin, C. (2018). IoT-based strawberry disease prediction system for smart farming. *Sensors 2018* 18 (4051): 1–17.

8 Ahmed, N., De, D., and Hussain, I. (2018). Internet of things (IoT) for smart precision agriculture and farming in rural areas. *IEEE Internet of Things Journal* 5 (6): 4890–4899.

9 Chen, W., Lin, Y., Lin, Y. et al. (2019). AgriTalk: IoT for precision soil farming of turmeric cultivation. *IEEE Internet of Things Journal* 6: 5209–5223.

**10** Chen, W., Lin, Y., Ng, F. et al. (2020). RiceTalk: rice blast detection using internet of things and artificial intelligence technologies. *IEEE Internet of Things Journal* 7 (2): 1001–1010.

**11** Hang, L., Ullah, I., and Kim, D.-H. (2020). A secure fish farm platform based on blockchain for agriculture data integrity. *Computers and Electronics in Agriculture*, Elsevier 170.

**12** Hyperledger. (2024). Available: https://www.hyperledger.org/resources/publications/walmart-case-study. 30th August.

**13** Lv, Z. and Singh, A.K. (2021). Big data analysis of internet of things system. *ACM Transactions on Internet Technology* 21 (2) Article 28.

**14** Sarker, I.H. (2022). Smart city data science: towards data-driven smart cities with open research issues. *Internet of Things* 19: 100528.

**15** Painuly S., Sharma S., and Matta P. (2021). Future trends and challenges in next generation smart application of 5G-IoT. In 2021 5th international conference on computing methodologies and communication (ICCMC) (8 April 2021) pp. 354-357. IEEE.

**16** Aggarwal, P.K., Jain, P., Mehta, J. et al. (2021). Machine learning, data mining, and big data analytics for 5G-enabled IoT. In: *Blockchain for 5G-Enabled IoT: The new wave for Industrial Automation*, 351–375. Springer.

**17** Ajmani, P. and Saigal, P. (2023). 5G and IoT for smart farming. In: *Applying Drone Technologies and Robotics for Agricultural Sustainability*, 124–139. IGI Global.

**18** Murugamani, C., Shitharth, S., Hemalatha, S. et al. (2022). Machine learning technique for precision agriculture applications in 5G-based internet of things. *Wireless Communications and Mobile Computing* 2022.

**19** Van der Aalst, W. (2016). Data science in action. In: *Process Mining*. Berlin, Heidelberg: Springer.

**20** Quiroz, I.A. and Alferez, G.H. (2020). Image recognition of legacy blueberries in a chilean smart farm through deep learning. *Computers and Electronics in Agriculture*, Elsevier 168.

**21** da Fonseca, E.P.R., Caldeira, E., Filho, H.S.R. et al. (2020). Agro 4.0: a data science-based information system for sustainable agroecosystem management. *Simulation Modelling Practice and Theory*, Elsevier 102: 102068.

**22** Tseng, F.H., Cho, H.H., and Wu, H.T. (2019). Applying big data for intelligent agriculture-based crop selection analysis. *IEEE Access* 7: 116965–116974.

# Index



*Current and Future Cellular Systems: Technologies, Applications, and Challenges*, First Edition.
Edited by Garima Chopra, Suhaib Ahmed, and Shalli Rani.

## e



## f